Reading Pound Reading

Reading Pound Reading

MODERNISM AFTER NIETZSCHE

Kathryne V. Lindberg

New York   Oxford

OXFORD UNIVERSITY PRESS

1987

Oxford University Press

Oxford   New York   Toronto
Delhi   Bombay   Calcutta   Madras   Karachi
Petaling Jaya   Singapore   Hong Kong   Tokyo
Nairobi   Dar es Salaam   Cape Town
Melbourne   Auckland

and associated companies in
Beirut   Berlin   Ibadan   Nicosia

Published by Oxford University Press, Inc.,
200 Madison Avenue, New York, New York 10016

Oxford is a registered trademark of Oxford University Press

Library of Congress Cataloging-in-Publication Data

Lindberg, Kathryne V.
Reading Pound reading.
Bibliography: p.
Includes index.
1. Pound, Ezra, 1885–1972—Knowledge—Literature.
2. Nietzsche, Friedrich Wilhelm, 1844–1900—Influence.
3. Modernism (Literature)   4. Criticism—United States—
History—20th century.   I. Title.
PS3531.082Z7384   1987   811'.52   86-23645
ISBN 0-19-504165-8 (alk. paper)

Permission to reprint the following material has kindly been granted by New Directions
Publishing Corporation:

William Carlos Williams, *Imaginations*. Copyright © 1970 by Florence H. Williams.

William Carlos Williams, *I Wanted to Write a Poem*. Copyright © 1958 by William
Carlos Williams.

Permission to reprint the following material has kindly been granted by New Directions
Publishing Corporation and Faber & Faber Ltd. (London):

Ezra Pound, *Cantos*. Copyright © 1934, 1937, 1940, 1948, 1956, 1959, 1962, 1963,
1966, 1968, 1972 by the Estate of Ezra Pound.

Ezra Pound, *Personae*. Copyright © 1926 by Ezra Pound.

Ezra Pound, *Collected Early Poems*. Copyright © 1926, 1935, 1954, 1965, 1967,
1971, 1976 by the Trustees of the Ezra Pound Literary Property Trust.

Ezra Pound, *Guide to Kulchur*. Copyright © 1970 by Ezra Pound. All rights reserved.

Ezra Pound, *Pavannes & Divagations*. Copyright © 1958 by Ezra Pound.

Previously unpublished material by Ezra Pound, copyright © 1986 by the Trustees of the
Ezra Pound Literary Property Trust, used by permission of New Directions Publishing
Corporation and Faber & Faber Ltd. agents.

2   4   6   8   10   9   7   5   3   1

Printed in the United States of America
on acid-free paper

This book is for
Mitchell Tanenbaum
of Brooklyn, if he
wants it

Preface

To begin at the beginning, by way of a prosthesis or manufactured substitute for the thesis developed over the course of this book, let me explain my title. The following chapters comprise a *reading* of Ezra Pound's *reading* procedure, a critique of his deployment of the sometimes quirky theories and practices of the various literary critics, philosophers, scientists, and even poets who, like Nietzsche, wrote in radically new or otherwise disruptive ways about culture and thus were incorporated into Pound's "Kulchur," a category he opposed to T. S. Eliot's primarily literary "tradition." This means that I deal almost exclusively with Pound's prose and primarily with those books and essays not directly concerned with poetry or his delineation of a modernist poetics. Careful reading of the rhetorical strategies of Pound's writings, the borrowed metaphors he used in place of method, and the attacks he launched against what he considered the (Eliotic) reaction against his own modernist revolution, involves the sort of "close reading" usually reserved for poetry.

This is not to say that my reading of Pound's reading is formalist or neo-New Critical. On the contrary, my approach to Pound's often neglected polemical and pedagogical writings not only refuses to subordinate the prose to the poetry, in the usual manner of supporting an interpretation of the latter by reference to the former, but it also challenges the mutual exclusivity of the creative and the analytic—of poetry as against criticism, art as against ideology. *The Cantos*, more a compendium of interpretations than a unified poem or an original epic, itself belies such easy oppositions, and, but for my resolve to avoid offering yet another reading of his *sui generis* work that might extend this study indefinitely, I might have spent more time on Pound's interpretative, translative, and

deeply intertextual verse. In turning briefly to the Pisan Cantos, Chapter
Five shows how Pound's poem reads his earlier poetry by using allusion,
quotation, and other questionable rhetorical devices to begin again and
yet "Make it New," to offer a style of avant-garde reading as or in place
of a belated (modernist) writing. Surely such palimpsestic interpretations
as Canto VII's placement of Henry James in Pound's revision of "The
Jolly Corner" ("And the great domed head, *con gli occhi onesti e tardi*/
Moves before me, phantom with weighted motion,/*Grave incessu*, drink-
ing the tone of things/And the old voice lifts itself/weaving an endless
sentence") are exemplary self-conscious reading scenes. Yet those and
many similar lines have received perhaps too much attention from com-
mentators on *The Cantos*. Indeed, that very passage, used to support an
apocryphal meeting of the two great American modernists, opens Hugh
Kenner's *The Pound Era*. But, while Chapter Two addresses Kenner's
powerful interpretations and his influential Pound-centered history of
literary modernism, I have for the most part deliberately avoided the
Poundians—especially since they tend to devalue the very texts important
to my assessment of Pound's own style of reading.

My approach does not yield a simple history of influence, of Nietzsche's
impact on Pound or of Pound's on his poetic cohorts and antagonists. In
this regard, let me explain what is at least the double register of my subtitle, specifically the ambiguity of its preposition. "After" in "Modernism
After Nietzsche" remarks a Nietzschean style of reading—that is a reading "in the manner of" as one might say of a painting after the school
or the style of a particular master. Quite apart from the chronological
accident that Pound wrote after Nietzsche, his literary and cultural criticism was informed by that German ironist's rigorous criticism of Wagnerian aesthetics and his assaults upon the genealogy of Western philosophy. In fact, as I show, Pound did not know Nietzsche well or even
directly. He got his Nietzsche, or his neo-Nietzscheanism, in translated
fragments mediated through the aestheticism and scientism of Wyndham
Lewis, Remy de Gourmont, Leo Frobenius, Oswald Spengler, and others. Even as he seems, perhaps unwittingly, to espouse a belief in Nietzschean irony (in destructive, if not deconstructive, thought), Pound provides a warning against the mapping of a Nietzschean influence. Take
for an example the following poetic fragment:

> I believe in double-edged thought
>   in careless destruction.
>
> I believe in some parts of Nietzsche,
>   I prefer to read him in sections;

> In my heart of hearts I suspect him
> of being the one modern christian;
> Take notice I have never read him
> except in English selections.
> (Redondillas)

More important than Pound's glancing references to, his misreadings of, Nietzsche is his preference for translations and fragments, a deliberate choice determining his treatment of poetic as well as critical precursors. This emphasis on the potentially distorting medium of language—and especially on polysemic metaphors and interested selections—is characteristic of Nietzsche's ironic and abusive interpretations of Wagner's grandiose operas and that composer-ideologue's dream of a composite, nationalistic art. Yet, as though conforming to the law—or falling heir to the fate—of reading that he did not know well enough either fully to break or strictly to enforce, Pound was both victim and erstwhile master of the fragmentation, the cross-discursive borrowings, the macaronic artistic and poetic languages, and the faded Classical inheritance that Nietzsche defined as modernism.

The nature of the texts under analysis and my mode of rhetorical or Nietzschean criticism dictate a certain degree of repetition along with a suspension of the thesis and redefinitions from different perspectives of certain key terms and concepts. Nevertheless, an outline of the chapters will indicate the logical progression of my argument. Chapter One defines the polemical sense of modernism that Nietzsche—and after him, though of necessity differently, Martin Heidegger, Theodor Adorno, Walter Benjamin, Michel Foucault, Jacques Derrida, their detractors and followers—discovered within and employed against the history of art and philosophy. For my purposes, Nietzsche's reading of Wagner and Heidegger's somewhat erroneous attribution of a nostalgic and metaphysical aesthetics to Nietzsche are most telling and economical; these provide the unsettling and recently critiqued philosophical background explicit in Chapters One and Two. In order to establish the parallels and intersections of Pound's various reading programs with Nietzschean interpretation, Chapter Two presents readings of those Poundian texts directly concerned with the methods and metaphorics of reading, such as "How to Read" and *ABC of Reading*. In those texts and elsewhere, Pound sought poetry and criticism that would be interpretive or interpretative. It is worthy of note that Pound used both words but not rigorously enough to maintain what would be a crucial distinction between an interpretive poetics, meaning one which offers a final version or explanation, and an interpretative poetics, or one which initiates an unstoppable process or infinite regress of interpretation.

Watch the use of these two terms (and other pairs like *totalized* and *totalitarian*, *tradition* and *orthodoxy*, *heteroclite* and *heterodox*), which Pound, as sceptical yet haphazard evader of system, quibbled over—sometimes deliberately. I have attempted to use and examine such terms rigorously throughout, though Pound, however suggestively, did not.

Chapter Three focuses on the Eliot-Pound relationship of the 1930s, the neglected and strained period of their friendship characterized by Eliot's increasingly literary and religious orthodoxy and Pound's contradictory adoption of lawless or heretical interpretations and his embrace of authoritarian politics. Here, the debate over *After Strange Gods*, conducted in fugitive little magazines and unpublished nasty letters, is important to an understanding of Pound's virtually deconstructive reading of the minutiae, hidden affiliations, and interested errors of Eliot's orthodoxy and his idea of tradition. Chapter Four's examination of Remy de Gourmont's impact on Pound's criticism, principally the strategy of dissociation and the notion of a spermatic reading economy, shows the strange centrality of that marginal Frenchman's writings to American modernism. Gourmont, who, writing of Herbert Spencer and Nietzsche, might have coined the verb "to deconstruct" (*deconstruir*), brought Pound closest to Nietzsche. This is evident in "The Translator's Postscript" to Pound's rendering of *The Natural Philosophy of Love*. Chapter Five, centered around *Guide to Kulchur*, demonstrates the mad search for method that brought Pound to Frobenius and Spengler, both of whom borrowed the rhetoric, if not the rigor, of Nietzsche.

It seems clear by the time of his writing *Guide to Kulchur* that Pound's reading procedure would have been impossible without Nietzsche and his interpreters. Likewise, current rhetorical and political reassessments of modernism, including the present one, would be impossible without the more recent and more informed translations, adaptations, and tropings—both continental and American—of Nietzsche. By way of considering these, and by way of opening onto a consideration of Pound's search for totalized art and his settling for a totalitarian state, Chapter Five closes with *The Cantos* and with the imperative to return to a direct and systematic treatment of Pound's politics which remains implicit here. This topic, which I have not avoided, requires further development in another—if not my next—book-length study.

My postscript, which traces Pound's reading of the American literary tradition back to Walt Whitman and Henry Adams, suggests a possible course for further study. Adopting the metaphor of the rhizome for the highly individualized and wayward tradition of American writing, I have shown the chronological and conceptual coincidence of Americanism and modernism. It seems that from its belated and much heralded beginnings in Emerson, American writing, never content to be simply or purely poetic, has been driven by the self-canceling impulse to be both originary and original. This has made for an exciting and often iconoclastic struggle of poets and readers, of poets who are also readers. Harold Bloom has brilliantly charted the combats of America's Orphic poets, thus belying earlier affirmations of American democratic individualism. But there is an even darker side to the American tradition, a side not unrelated to Pound's Nietzschean—but also Fascist—reveling in the individual, the active, the disruptive, and the metaphoric. It is this strain of Americanism, as present in Whitman as it is in Pound (perhaps, as Pound claimed, as much in Jefferson as in Mussolini), that at once resembles rhizomatic (that is, weedlike, asystematic, antihierarchical) thinking and must be analyzed by it. Nietzsche, whose own writings have been adopted by reactionaries as often as by revolutionaries, cautions against falling into the (Poundian) trap of enshrining a new god of interpretative free play in place of the old gods of systematic and cultural hegemony. I will end my prefatory remarks with a rather lengthy quotation that stands as a warning and promise of what comes of reading Pound, of recognizing another of the all too frequent marriages of avant-garde aesthetics and reactionary thought resulting from taking metaphors too seriously or not seriously enough. Therefore, I will allow Nietzsche the last first word:

> But I should think that today we are at least far from that ridiculous immodesty that would be involved in decreeing from our corner that perspectives are permitted only from this corner. Rather has the world become "infinite" for us all over again, inasmuch as we cannot reject the possibility that *it may include infinite interpretations*. Once more we are seized by a great shudder; but who will feel inclined to deify again after the old manner this monster of an unknown world? And to worship the unknown henceforth as "the Unknown One"? Alas, too many *ungodly* possibilities of interpretation are included in the unknown, too much devilry, stupidity, and foolishness of interpretation—even our own human all too human folly, which we know.

Acknowledgments

I owe special debts of gratitude to Joseph N. Riddel, the generous friend who first directed me to the study of modernism; to John Espey, who taught me the value of Pound's imperfections; to Murray Krieger, who introduced me to the formal study of literary criticism in my first graduate seminar; to the late Eugenio Donato, whose kindness and intellectual honesty I remember; to Shuhsi Kao, a better model than I can hope to follow; and to John Mardirosian—a more patient and entertaining research assistant is unimaginable.

Not to exaggerate the pieties of the occasion, may the following friends and colleagues be blessed for their help and absolved of responsibility for my mistakes and omissions: Sacvan Bercovitch, Warner Berthoff, Linda Brooks, Eduardo Cadava, Jules Chametzky, Dorothy Clark, John Espey, Eugenia Gunner, Miriam Hansen, Robert Kiely, Jon Klancher, Joseph Kronick, Stephanie Lagoy, Tejaswini Niranjina, David Perkins, Geneva Phillips, Joel Porte, John Rowe, Adelaide Russo, Paul Sheats, Ross Shideler, Werner Sollors, Maureen Spofford, and Melissa Weissberg.

A grant from the Hyder E. Rollins fund of Harvard University aided the publication of this book. The Henry E. Huntington Library and the University of California at Los Angeles supported my research with study and travel grants. I thank the librarians and staff members at the following libraries: the Beinecke Rare Book and Manuscript Library, Yale University; Rare Book Collection, Regenstein Library, University of Chicago; Special Collections at Cornell University Library; Bancroft Library, University of California, Berkeley; Houghton Library, Harvard University; Humanities Research Center, University of Texas, Austin; the Poetry Collection of Lockwood Library, State University of New York at Buffalo.

An earlier and shorter version of Chapter Four appeared in *boundary 2* (Spring/Fall 1984)

Contents

Reading Pound Reading

CHAPTER ONE

Reading, Writing, and Rhetoric
Nietzschean Traces in
Pound's Interpretive Poetics

> how shall philologers?
> A butcher block for biographers,
>   quidity!
> Have they heard of it?
> 'Oh you,' as Dante says
>   'in the dinghy astern there'
>   Canto XCIII: 631[1]

When I imagine a perfect reader, he always turns into a monster of courage and curiosity; moreover, supple, cunning, cautious; a born adventurer and discoverer. In the end, I could not say better to whom alone I am speaking at bottom than Zarathustra said it: to whom alone will he relate his riddle? "To you, the bold searchers, re-searchers, and whoever embarks with cunning sails on terrible seas— to you drunk with riddles, glad of the twilight, whose soul flutes lure astray to every whirlpool, because you do not want to grope along a thread with a cowardly hand; and where you can *guess*, you hate to *deduce*."

NIETZSCHE, *Ecce Homo*[2]

If we want to get as close as we can to Nietzsche's conception of the will, and stay close to it, we are well advised to hold all the usual terminology at a distance. Whether we call his conception idealistic or non-idealistic, emotional or biological, rational or irra-tional, in each case it is a falsification.

HEIDEGGER, *Nietzsche*[3]

How can one justify this odd coupling of the names Pound and Nietzsche, of poetry and philosophy, especially in a book that rejects the old struc-ture of influence? Here, at the outset, let me state that I will not argue a continuous filiation between the two writers or two disciplines, not even

3

the affiliation of a poststructuralist "New Nietzsche" and a postmodern Pound.[4] Pound's and Nietzsche's texts do not delineate a tradition or an orderly genealogy of works and ideas either. Nevertheless, their differences do not prevent comparison of the surprisingly like strategies they employ. But such equations and juxtapositions have little to do with literary history in the traditional sense and will be less satisfactory to historians than to critics. It is, therefore, with the question of interpretation each opens up within his respective discipline, a question which blurs the classical distinction between poetry and philosophy, that I begin. In view of the palimpsest of readings layering Pound's texts, the apparently infinite regress of allusion, citation, quotation, and the like—the "wash upon wash of classicism"—that characterizes both the *Cantos* and the prose, we must be as self-conscious about reading Pound as recent rhetorical critics have been about reading Nietzsche.

Clearly, Pound, who went so far as to equate poetry with interpretation (the best poetry being, like the Troubadour's Mystery Cults, "interpretative," SR, 87), obsessively sought one method which might yield the right idea of tradition. And, as time went on, his criticism was written with proper readings of his own poetry in mind. Yet he arrived at several conflicting notions and metaphors of reading. These were hardly a "method," even if he did seem to propose in his early criticism a "New Method of Scholarship." His criticism and poetry consist of, indeed thrive upon, conflicting interpretations. And, like Nietzsche, Pound would sometimes argue for such interpretive free-for-all—even rhetorical "free-play." In his own regional discourse (dare we call it poetry?), Pound happened upon the same problematics of language that inform Nietzsche's more studied—and playful—interrogations of art and (philosophical) method.

Chapters Four and Five will treat thoroughly those Poundian texts where one might expect to encounter a theory of reading: "How to Read," *ABC of Reading,* and *Guide to Kulchur.* In this chapter I will enquire more generally into the status of Pound's criticism. Taken as a whole, or even carefully examined piecemeal, his criticism hardly presents a coherent aesthetics on which to base readings of works in the tradition or to found an "idea" of "the modern." Indeed, the very possibility of hermeneutical mastery propounded by systematic aesthetics is everywhere put into question by his notion of reading, which is decidedly against "ideas" and markedly resistant to generalizations identified as "modernism." This is not due to Pound's refusal to engage ongoing arguments about the various arts or to strive for a comprehensive aesthetics. On the contrary, he wrote extensively on music, literature, and the visual arts; he even included biology, anthropology, mathematics, and physics in various schemes to translate art and science into a workable tradition, explicitly to read the various parts of culture (Kulchur) as a whole. Rather than comprising a system, however, his occasional criticism is only now being issued in separate books, as if under clearly delineated topics and theses (*Ezra Pound on the Visual Arts* and *Ezra Pound on Music,* for example), which testify instead to his inability to master all genres in a totalized reading. Just so, the heterogeneous nature of his critical texts belies our own urges to keep readings of his poetry and criticism within proper bounds (or books).

One cannot find a single master thesis in the fragments, allusions, ellipses, clichés, repetitions, catachreses, and self-(mis-)quotations that comprise his reading program. We might, however, discover something more pertinent to the questions that recent criticism or, more exactly, recent theories of reading have posed as a question of modernist writing and/or reading. At the very least, one discovers that Pound rigorously—if sometimes accidentally—challenges our traditional habits of reading. His notion of reading is not simple. In fact, whenever he begins to articulate his style of reading he seems compelled to pose the most radical questions: "What is literature?" (ABCR, 28), "Why Books?" ("How to Read," LE, 20), "What is language?" (ABCR, 28). These questions, as well as their provisional answers, which lead him to affirm, among other things, that all language is originally figurative and interpretative, disturb the aesthetic and political hierarchies he ostensibly fought to maintain. Thus, he repeatedly makes discoveries that he did not intend and could not fully appreciate.

We must recognize that the disorganization, the maddening inter- (and intra-) textuality of Pound's criticism, is not wholly fortuitous. His use of fragments and his attention to irreducible—if "luminous"—details appear to the contrary a critical or textual strategy. His criticism, part of a lengthy, heated polemic against all takers, is not merely *a*systematic. While I do not mean to reverse the usual devaluation of his prose in relation to his poetry, one must recognize that his notion of reading is often stunningly *anti*systematic. At first blush, then, it might seem odd to treat Pound's works with those of a German philosopher (or, to use Pound's word, "flyosopher," letter to T. S. Eliot, January 1940). But who more than Nietzsche explored the modern desire to comprehend all culture in a totalized reading, condemning in the same gesture the very tactics he was forced to employ? If Nietzsche was unusually conscious of the implications of his procedure, Pound was only intermittently aware of the force of his own disruptions, or "instigations" as he sometimes called them. His reading *theory/practice* (one of several oppositions difficult to maintain with regard to Pound) represents an uncanny coincidence of philosophical innocence with a pronounced knack for uncovering the embarrassing contradictions in the grounding assumptions of Western aesthetics and the metaphysical tradition at large.

The most remarkable examples of Pound's insights into the disruptive potential of language occur in *Guide to Kulchur*. Most of these discoveries—or disclosures—take the form of apologies, ironic asides, and seemingly gratuitous polemics that nearly efface Pound's seriousness as an author and certainly undermine any possibility of detached or scientific authority. Had Pound been a systematic thinker (the "philosopher" of pre-Nietzschean times), this last "textbook" on reading might have defined his method, extending it through new and clearer examples. Instead, his final words on reading are a loose collection of his earlier programmatic statements, a series of notes toward a new(er) method to come, and an apparently random catalogue of favorite quotations and memories. Although issued as a "book," *Guide to Kulchur* is clearly a "text" to be reckoned with.[5] The title of the first American edition of this book, *Culture: The Intellectual Autobiography of a Poet,* makes clear his traditionally serious purpose. But that title, if not its seriousness, was assigned by American editors, who have always been leery of misspellings like "Kulchur." Editorial intentionality aside, *Guide to Kulchur*'s simple "loose leaf" format nearly requires rewriting at every reading, because it causes one to reassess Pound's earliest notions of poetry and culture in light of new revisions—and vice versa. In fact it is nearly as aleatory and uncontainable as, say, Nietzsche's posthumous "book," *Will to Power* (*Wille zur Macht*), those surviving fragments gathered into various books by Frau Förster-Nietzsche and a continuing series of interested readers and/or editors, running the gamut from the National Socialists to contemporary Marxists who would reappropriate Nietzsche for a different revolution.

Perhaps Nietzsche would have separated *The Will to Power* into smaller, more manageable works; after all, long passages of it found their way into those earlier volumes he saw through publication. But Pound had no such plans. Instead, he insisted upon the heterogeneity of his last reading text, even if he viewed it as part of a series:

> I am, I trust patently, in this book doing something different from what I attempted in *How to Read* or in the *ABC of Reading*. There I was avowedly trying to establish a series or set of *measures, standards,* voltometers, here I am dealing with a heteroclite series of *impressions*. (GK, 208; emphasis added)

The closer one comes to Pound's reading texts, the more problematic become clear oppositions like "standards" and "impressions," theory and practice, generalizations and details—or the whole series of oppositions that might be abbreviated as "(our) reading from (his) writing."

Furthermore, when one begins to interrogate his notion of "reading," there immediately arises the question of his poems which also consist of translations and interpretations. All are readings which disrupt both the formal and rhetorical traditions of the poetic or of "poetic language." Careful attention to the details of his most marginal critical works will thus subtly disturb the privilege he affords first to his own and then to all poetry.[6] One cannot easily return to traditional and privileged definitions or simply accept those tempting critical apothegms for which Pound is famous. All this complexity ensues despite his frequent protests that he would enclose his ideas, along with all of tradition, in clear definitions and appropriate forms. Addressing his quest for both the adequate word and the proper critical genre, he says: "I am trying to get a *bracket for one kind of ideas,* I mean that I will hold a whole set of ideas and keep them apart from another set" (GK, 29; emphasis added). He thus casually assumes the logic of identity upon which philosophy—and all precise language—depends. Yet this same principle of identity, the effort to assign a clear and stable meaning to each word, he everywhere subverts, or at least puts in question.

We begin to understand the importance of that exemplary case of Pound's concern for system or taxonomy by pausing over his italicized word *"bracket."* He could not have known the use to which modern philosophy would put this simple notion, turning it into a master concept, whether in Husserl's eidetic reduction or Heidegger's re-marking of effaced concepts. Nevertheless, his dogged attention to certain key categories resembles these later developments in the critical reading procedure that Nietzsche inaugurated. For example, Heidegger places *brackets* around the abused, overused, or neglected philosophical terms he is nevertheless forced to employ in his interrogations of philosophy. By literally framing such problematic words within brackets, he dislodges them from their position as the ordinarily invisible "enframing" (*Gestellen*) or defining concepts of metaphysics. Such concepts ground all definitions and are so fundamental that they seem to require no definition or interrogation, though they conceal many assumptions and potential linguistic disturbances.[7] Heidegger's most famous brackets are those around "Being," a word worn down and overessentialized by the very philosophical tradition he wants to redirect and revitalize. Yet he has no choice but to use this word, even in his attempt to isolate it from his preferred words and notions. Though he leaves "being" aside, Jacques Derrida's "erasure," which is at once an italicizing and an obliteration of even his own too philosophical coinages, is a yet more violent way of marking necessary but inadequate words. Derrida hereby takes Heideggerian bracketing one step further and into a "subtle parody" (his phrase for Nietzschean reading) of the necessarily incomplete interrogations of philosophical language. In part because he focuses on the inadequacies of his own language and style, Derrida, whose "erasure" becomes a self-critique and almost literally a self-effacement, arrives at a style of interpretation (an "interpretation of interpretation") that is more rigorous in a peculiarly Nietzschean (ironic) way than that of his precursors, including Husserl, Heidegger, and Nietzsche. As we shall come to see, Pound innocently happened upon a notion of reading potentially as rigorous and every bit as disruptive as that of Derrida.

In the interest of clarity and economy, let me give an abbreviated account of such (*deconstructive*) readings, using the notion of bracketing or erasure and Derrida's now familiar final step of "reinscription." First, a grounding concept is wrenched from its context in a philosophical argument or aesthetic work and placed within brackets. Next, its assumptions, affiliations, implications, and conventional meanings are exposed and dissected. Third—and this is where most readings complacently stop— the term thus devalued is returned to its discourse revalued with a new and narrower if not altogether precise meaning (or better, *meaning-function*) which marks itself as a metaphoric and undecidable nonconcept. One should note that these three steps do not proceed with dialectical inevitability, but instead by disrupting dialectics. In a rigorous (but also a playful and self-destructive or "Nietzschean") reading, the bracketed term does not make its way back intact. Instead, when returned, or misreappropriated, into the reader's discourse, it is marked (and re-marked nearly to the point of effacement or "erasure") with more meanings and potential interpretations than it previously had, more meanings than it can bear. Such a reading supplements, rather than rejects, the discourse it inherits, raising more questions than it purports to answer. Thus, and one must interrogate intentionality here, Heidegger's "question of Being," which should be printed and spelled "[Question] of [Being]," remains open despite his efforts to found a new philosophy on it. Furthermore, several other concepts, questions, and metaphors can no longer be read as innocent, as free of the distortions to which they have been (and Nietzsche suggested "should be") subjected by Nietzschean readings. Among the more familiar words thus suspect are "reading," "writing," and "rhetoric," all of which are involved in Nietzsche's idea—is it a concept or a strategy?—of overcoming (both *Verwindung* and *Überwindung*), itself an important metaphor for modern reading and writing.

One can list many similar Poundian metaphors for metaphor—metaphors, that is, for the reading act. "Luminous Detail," "Vortex," "Image," "Paideuma," "kinematic art," "palimpsest," "interpretative metaphor" are all indirect descriptions of the exchanges constituting the rather idiosyncratic "living tradition" he traced from the pre-Socratics to his own readings. Suffice it to say that Pound's hoped-for "bracket for one set of ideas" is no light matter, that his own self-reflexive and subversive reading program frequently suggests the manner in which more philosophically sophisticated writers have *read*—both created and criticized or "destructed"—their traditions and inherited methods. If Pound did not *bracket* problematic or abused terms with the aim of rigorously redefining and stabilizing them as instruments of his "New Method of Scholarship," for most of his career he waged guerrilla war against the terms and texts of the Eliotic tradition. Rather than erect a countertradition of alternative works under the category "literature" or "tradition" or "culture," he continually returned to those "luminous details" first presented in "I Gather the Limbs of Osiris." He privileged canonically marginal and metaphorically or interpretively dense passages in Homer, Dante, the Troubadours, and the occasional writings of various historical figures. For Pound, these were vortices of energy and change, not simply representatives of a stable tradition. By the time he turned from his various "ABC's" and "How To's" to make a general statement about culture—"Kulchur" or just plain "Kulch"—he had gathered a bundle of his own and others' conflicting readings. And, more disturbingly, from a variety of readers—many of whom, like Burckhardt, Gourmont, Frobenius, and Spengler, had been touched by Nietzsche—he had borrowed exotic names for the process he wanted to extract from his readings and turn into a method. To this loose-leaf thesaurus of texts and readings he gave the old name Culture. Well, not exactly: he marked the old category with his abusive spelling, Kulchur, though he could not fully efface the expectations one generally has for such a comprehensive notion of those traditions which include the literary Eliotic tradition and, more urgently, the threat of German political and cultural hegemony, which had traveled since before Wagner under the name *Kultur*. Pound's revisionary translation of this German category will receive full treatment in Chapter Five.

We cannot read modern "culture"—as word, concept, or metaphor—in quite the same way after Pound, who, with a characteristically uneasy balance of reverence and irony, changed its meaning with its spelling to Kulchur.[8] His "misspelled" title, a word that recurs in his other critiques of too narrow or static definitions of tradition, has a meaning that approximates Frobenius's *Kulturmorphologie,* which is a disruptive or kinematic mapping of cultural forms. Kulchur marks an aberration in the classical idea of culture as received ideas, even if we take his spelling to be a lapse of memory or a refusal to accept the simplest rules. His most familiar definition of culture suggests both the traditional sense and more: "Knowledge is NOT culture. The domain of culture begins when one HAS 'forgotten-what-book'" (GK, 134).[9] Pound's "Kulchur" is made up of "current" events and residual "ideas" from disparate times and places, not all of which find critical or academic favor. Containing as it does random, even forgotten elements, it differs fundamentally from Culture or the (Eliotic) tradition, understood as the inherited memorial to the great accomplishments (almost "The Great Books") of Western civilization. Little wonder that readers with their own designs on saving culture intact might wish to forget (with a Nietzschean, purposeful, forgetting) the details of Pound's facile demolition of conventional culture.

With similar ease, and an irony that can only be termed *dramatic* (in at least two senses, since Pound is the butt of his own joke), Pound annuls the ideal of a word which is adequate to the thing-in-itself and therefore to the sort of "Truth" (self-identity, Presence) acceptable to Western philosophical and literary tradition(s). Since before Plato, philosophy has insisted on a Truth prior to language. As though in defense—and literally in a series of Defenses of Poetry—literature has made the collateral assertion that poetry is based on universals unchanged by time and translation, albeit on a poetic or figurative truth yet one immune from political, social, economic, or historical forces. Eliot's Anglo-American modernism, that body of new and conservative readings of inherited works, is a most potent defense of poetry from the scruples of abstract philosophy and the perhaps greater threat of historicity. But Pound endorses neither defensive maneuver; for him, poetry is the highest cultural form precisely because its figurative language both assaults philosophical truth and will not be categorically contained.

The more leisurely investigations that follow in subsequent chapters will justify my economy in drawing so much from a few hints, but here we should simply remind ourselves that Pound uncovered something quite "different"—another loaded word in "modern" philosophy after Nietzsche, Heidegger, and Derrida—from the originary poetry, the grounding word he ostensibly sought in writers ranging from Heraclitus to William Carlos Williams. When he inquired into the use and source of language, instead of *Logos, the Word,* he found words originating in a disseminating force, even when he made fond reference to the conceptual power of *logos* in its earliest Classical sense. Such words, plural as well as figurative, can scarcely be made to recover, if they ever possessed, stable meaning. Pound offers the following (de-)definition he claims to have read in Heraclitus: "'From god the creative fire, went forth spermatic *logoi,* which are a gradual and organic distribution of an unique and spermatic word (logos)'" (GK, 128).[10] Far from being stable, let alone comprising the grounding principle of identity, this first ("spermatic") word is at best a temporary containment of difference. In this way, Pound virtually anticipates any extreme to which Derrida would force "différance" and "dissemination." A poet known for his hostility to—not to say his sheer ignorance of—philosophy would seem to undo onto-theology in its own terms. He still held on to the belief that poetic language might penetrate to the essence of things. We will come to recognize these moments of extreme systematic tension or insightful blindness (to turn slightly Paul de Man's trope for modernity) as fundamental to Pound's readings and to his own modernism, which is as much a way of reading tradition as it is a corpus of modern writing.

II

It is no secret that Pound's generation ("The Men of 1914") sought membership in the artistic aristocracy of *Übermensch,* that such themes as "eternal recurrence," overcoming, and "*gai scienza*" appealed to those humanists engaged in anthropological (even archeological) reconstructions of Western civilization. Especially during the years immediately preceding World War I, when he frequently contributed essays and translations to *The New Age,* A. R. Orage's little magazine for popularizing Nietzsche and other continental fads, Pound was exposed to all sorts of Nietzscheanisms. Obvious connections exist between (readings of) Nietzsche and Pound. For example, one can trace the ambiguous heroism of Sigismundo Malatesta to Wyndham Lewis's Machiavellianism, an "ethical transvaluation" of Nietzsche.[11] And one is justified in saying that Remy de Gourmont's rhetorical strategy of "dissociation," "transvaluation of values" by another name, had a significant and sustained impact on Pound's redefinitions of key aesthetic notions within the tradition.[12] Much later, of course, Pound was to borrow the Spenglerian notions of *paideuma* and *Kultur* through Frobenius; these too have been traced back to Nietzsche. On such evidence one might erect quite formidable historical or thematic readings of the criticism in order to claim various Nietzschean influences on Pound.

In any event, it is easy to discount Pound's own underestimation of Nietzsche when, for example, he dismisses that philosopher as a "temporary commotion," a distraction from his own announced quest for words with precise definition and a poetics that might compete with science and other progressive discourses. In "How to Read," he acknowledges Nietzsche's disruptive potential but does not lay stress on the critical force of reading, let alone on the neo-Nietzschean critique of modern discourse:

> Thought was churched up by Darwin, by science, by industrial machines, Nietzsche made a temporary commotion, but these things are extraneous to our subject, which is the *art of getting meaning into words*. (LE, 32)

Along the lines of literary historical influence studies, one might further argue that Nietzsche was a large part of Pound's famous "nineteenth-century inheritance," the post-Symbolism he acquired second- and third-hand mainly through Gourmont. Pound himself offers evidence that he might have spared Nietzsche—who was always thought to be a rather poetic or simply an asystematic philosopher—the harsh treatment he usually affords German and other philosophers. Speaking of Gourmont and perhaps alluding to the commonplace that Nietzsche was responsible for World War I, he says: "Nietzsche has done no harm in France because France has understood that thought can exist apart from action" (SP, 421).[13] Such statements permit one to trace a line of deflected influence from Nietzsche and Wagner through French Symbolism to Pound. Remember, too, Nietzsche's own assertions that Baudelaire was "the first intelligent reader of Wagner anywhere" (EH, 248). Nietzsche was at least the formidable second, borrowing Wagner back from poetry to give his music rhetorical power over philosophy.

Such readings are tempting, though they would read like a pedant's dream, a hermeneut's nightmare, or a typical interpretation of the meaning of any of Pound's Cantos. In order to execute such a project by a chain of privileged quotations, one would have to hypostatize Pound's— not to mention Nietzsche's—main ideas and make them part of a stable canon within (or against) "the modern tradition." All this while assuming that there are complete, authorized texts to work from—which is to indulge the historian's illusion of a univocal tradition or the textualist's illusion of an authorized version. Such assumptions are precisely what Pound's notion of reading, and his individual readings, challenge. He shares his "ab-use" of historicism with Nietzsche; his late criticism also came to share that German critic's refusal to separate thought from action or simply to endorse the old discursive categories and procedures. Therefore, we must look more carefully at what it means to read (like) Nietzsche, which is a habit transmitted quite differently from the passing on of literary traditions in the Arnoldian or Eliotic mode. Pound encountered Nietzsche, directly and indirectly, as he ranged from avant-garde art to French letters to German *Kulturmorphologie*. But such close encounters—of an intertextual rather than a textual kind—are less important than the fundamental properties of reading and writing, the instability and figural energy of language that Pound and Nietzsche stressed in modern art and/or criticism.

Nietzsche's critique marked neither the end of philosophy nor the reification of poetry and art, but a brilliant staging of the perennial burden of those who deal in language. In various rhetorical games played against philosophical grammar, he affirms the inherent rhetoricity of language which logic would tame or, better, ignore. Following Paul de Man's definition of Nietzsche's rather complex notion of the two types of rhetoric, one might say that Nietzsche affirms the necessary figuration (rhetorical tropes) of words which unsettles any rhetorical position or argument. Nietzsche's readings uncover the complicity of art and philosophy and their mutual dependence on opposed metaphysical concepts—such as identity and difference, theory and practice, sensible and intelligible, necessary and contingent—which can be sustained only in the guarded and privileged language of philosophical constructs. This is to say that he exposes the repressed desires, assumptions, and virtually uncontainable tropology behind the fundamental tools—concepts, works, schools, and systems—of the philosophical dialogue which apparently continued uninterrupted from Plato through Hegel. By focusing on accidents of language and embarrassing confessions, he shows that Western philosophy is a discontinuous effort to idealize consciousness and banish contingency, or to privilege philosophical Ideas over mere words and over the dangerous sensuousness and sensuality of art. Commenting on his engagement of what to many seemed the sterile world of ideas in the abstract realm of metaphysics, he insistently bombards established principles and goals (the "good" of religion and the Truth of German idealism) with a variety of rhetorical strategies that traditionally had been banished from pure "Philosophy":

> Philosophy, as I have so far understood and lived it, means living voluntarily among ice and high mountains—seeking out everything strange and questionable in existence, everything so far placed under a ban by morality. Long experience, acquired in the course of such wanderings *in what is forbidden,* taught me to regard the causes that so far have prompted moralizing and idealizing in a very different light from what may seem desirable: the *hidden* history of the philosophers, the psychology of the great names, came to light for me. (EH, 218)

Here Nietzsche would seem to practice the sort of heroic reading defined, in this chapter's second epigraph, as a monstrous curiosity. He undoes the impersonality of philosophy by disclosing the equation of morality and metaphysics throughout the history of philosophy and in that art which aspires to philosophical respectability.

In this regard, Nietzsche's diagnosis of Wagner's "modernism" and the Wagnerian ideal of *das Gesamtkunstwerk,* as an encyclopedic glorification of German thought and *Kultur,* is of primary importance. *The Case of Wagner* (*Der Fall Wagner*), with its titular pun on Wagner as a medical case and an exemplary modernist, explicitly poses Nietzsche's infamous "question of style" and requires careful consideration. To begin, I would underscore what de Man has noted in Nietzsche's general style of reading that forces so many interpretations to avoid or miss the rigor it demands:

> If we read Nietzsche with the rhetorical awareness provided by his own theory of rhetoric we find that the general structure of *his work resembles the endlessly repeated gesture of the artist "who does not learn from experience and always again falls in the same trap."* What seems to be most difficult to admit is that *this allegory of errors is the very model of philosophical rigor.*[14] (emphasis added)

Here, necessary "errors" might be opposed to accidental or possible "mistakes," which arise through carelessness or ignorance. Yet neither errors nor mistakes are always avoidable or clearly distinguishable. According to de Man's own gloss, "cognition cannot be separated from discourse. But, if error is thus 'fundamentally' linked to cognition, with all the indeterminacy inherent in this metaphor of foundation, then the mind cannot be expected to master the distinction between possibility and necessity."[15] We will again have occasion to reflect upon the distinction between error and rigor in questioning the foundations of modernism and the attempts of modern artists to correct various mistakes of their forebears.

By radically questioning philosophy's apparent errors and readjustments of the system from the pre-Socratics to Hegel, by reading philosophy's dead metaphors through a series of more disruptive tropes, Nietzsche works his strategy of affirmative—or, in a word, deconstructive—reading. He is not so much after an error-free system as a symptomology that reads through mistakes to uncover the erroneous separation of art and rhetoric from systematic philosophy. But not even Nietzsche, perhaps the craftiest of all rhetors, completely masters his own errors. Instead, his texts inscribe a reading style that would seem to celebrate its own "vertiginous non-mastery," Derrida's phrase for the ambiguous control Nietzsche exercises over his own interpretive metaphors.[16]

Now, especially after Derrida's and de Man's readings of Nietzsche, we could scarcely want to go directly to Nietzsche's text, let alone Pound's. Freedom from intertextuality is impossible; instead, it is necessary to consider the curious fate that befell Nietzsche at the hands of academic philosophers—a fate not unlike Pound's at the hands of traditional literary critics. Here Heidegger's attempt to punctuate the philosophical canon with a reading of Nietzsche's Will is worth considering. Heidegger was the first to interrogate all Western philosophy (which for him is represented by ontology) in the figure of Nietzsche, the first to take "transvaluation" seriously. His reading remains the most rigorous (yet at the same time the most deliberately blind and metaphorically "errant") attempt to bring Nietzsche within the boundaries of philosophical discourse, characterizing that precursor as the consummate ontologist who nevertheless opens a way of redefining Being. Derrida, himself still in the throes of Heidegger's infectious reading style, must admit:

> Rather than protect Nietzsche from the Heideggerian reading we should perhaps offer him up to it completely, underwriting that interpretation without reserve and up to a certain point . . . where the text finally invokes a type of reading more faithful to his [Nietzsche's] type of writing.[17]

While we cannot pretend to such unreserved rigor, it is possible to recognize the perils and rewards of reading Nietzsche *into* and thereby against the metaphysical tradition. Inserting Nietzsche's strategic disruptions into traditional philosophical inquiry, to which literary criticism and other disciplines have looked for Truth, uncovers a repressed history of philosophy's attempts to gain hegemony over art. In order to bring intepretation within proper bounds and hence to define the philosophical canon and canonical readings, philosophy has repressed figurative language. We should not, however, exclude Nietzsche from philosophy or protect him from Heidegger's interested repetition of the gesture of canonization that has always characterized serious thought. Likewise, we should not abandon traditional literary methods or attempt to ignore the Anglo-American prejudices that have helped to canonize Pound's early poetry and dismiss his Nietzschean style of reading.

As Heidegger shows, Nietzsche's highly metaphorical style is the greatest impediment to his incorporation into the philosophical canon. Of this Nietzsche was fully aware, for style is one of the things philosophers take for granted and that he would therefore expose. Nietzsche demonstrates that, excluding such notable cases as Plato and Schopenhauer, philosophers have tended to eschew rhetoric and a poetic style for clear, logical arguments—much in the same way that nineteenth-century science turned from metaphysics to logic and mathematics for an adequate language.

Philosophers have always sought precise definitions and comprehensive statements, carefully avoiding rhetorical figures more appropriate to dogmatism and art. Metaphors and other figures nevertheless might occasionally be borrowed to convey otherwise inaccessible abstractions. In other words, philosophers sometimes risk one sort of rhetoric (tropology) to ensure success at the other (persuasion), all the time believing that their own language could exchange ideas free from the distortions of metaphor. Nietzsche, on the other hand, reveals in an unphilosophical style, a hybridization (some might say bastardization) of poetic styles and philosophical Ideas. Not only does Nietzsche's remarkably stylized German prose attest to his mastery over the empirics of style (which cannot be treated here, in an American[ist] book), but it also exposes stylistic play in the Western philosophical tradition which has always tried to exclude such mistakes. The following description of his own prose, at once penetrating and obfuscating, dares readers, always proud of the difficulty and rigor of philosophy, to adopt Nietzsche's interpretative (and disorienting) style:

> My writings are difficult; I hope this is not considered an objection? To understand the most *abbreviated* language ever spoken by a philosopher— and also the one poorest in formulas, most alive, most artistic—one must follow the *opposite* procedure of that generally required by philosophical literature. Usually, one must *condense,* or upset one's digestion; I have to be diluted, liquefied, mixed with water, else one upsets one's digestion. (EH, 340)

Nietzsche only seems to endorse attempts to essentialize his argument, to turn or tame programmatic aphorisms into "critical questions." A long and twisted history of philosophical and political interpretations of Nietzsche's key metaphors testifies to the risks inherent in reading such writings that will not stay safely within the bounds of one discipline or the bindings of coherent "books." His texts keep growing; never stable, they change with every reading in ways more obvious than most philosophical writings. Mere fragments of Nietzsche's mad scribblings have propagated volumes of interpretation, just as he returned to expose apparently stable concepts to interpretive free-play. Most recently, Derrida has executed a parody of textualism, psychoanalysis, phenomenology, and structuralism by a reading of the last note Nietzsche is thought to have written: "I have forgotten my umbrella."[18] Leaving untouched the Freudian and other Franco-Germanic suggestions of such "unreadable" fragments, we are reminded of the condition of Pound studies. As papers steadily emerge from trunks at Yale's Beinecke Library and elsewhere, drafts and fragments (and "Drafts and Fragments") are still being added to *The Cantos* and other more obscure parts of Pound's corpus, which might yet accommodate Nietzsche—if not his umbrella or styl(us).[19]

III

We must recognize that for Nietzsche style, both in art and philosophy, is virtually inseparable from wider questions of language and interpretation. He never contented himself with formalism or with a Kantian idealization of art; neither did he subjugate stylistics to the traditional concerns of history or politics. Instead, and this is part of the disturbance Heidegger and others would dismiss, he forces the customary categorical oppositions upon which aesthetics depends to tremble and merge. Against idealism, he juxtaposes a physiology of art. Against "serious" historicism, he seems to argue frivolous details. Yet his arguments are not simple reversals of traditional hierarchies; even less do they promise resolutions to the old dilemmas. Instead, they force us to examine the form and assumptions of artistic production and philosophical argument.[20] This double reading is most fully elaborated in Nietzsche's critique of the Wagner cult at Bayreuth. By an ever sharper focus on the Bayreuth performances, as Wagner's attempt to persuade his audience to an ideal (mythos) of German nationalism and primitivism, Nietzsche attempts to overcome Wagner's romanticism only to affirm the more questionable virtues of modern decadence. He uses Wagner to overcome Wagnerism.

At this point in Nietzsche's text, questions arise which disturb traditional notions of influence and interpretation: what did Nietzsche mean by "overcoming"? What did he hope to accomplish by deploying this notion or procedure? It is important to note that he used two different words translated into English by the one word "overcoming." The dissonance in meaning between the two German words is crucial to Nietzsche's notion and practice of reading. Recognizing this dissonance does not resolve the complexity of such readings; it only adds to their interest (an "increment of association," as Pound would say). Nietzsche's two words are *Überwindung,* literally "overturning," an overcoming in the sense of conquering and going beyond; and *Verwindung,* an archaic word which means torquing or troping, in the sense of a repetition that revises while preserving the original.[21] The differing senses of these words offer a reading of the Hegelian *Aufhebung,* which means to transcend without rejecting, to sublate or even incorporate. Nietzsche's revision of Hegel's term rests on the notion of *Windung* (contained in both his words); that is, "turning" in the sense of linguistic troping and in the sense of the perpetual revolution of meaning against the (Hegelian) resolution of the dialectic. Nietzsche positions this revolutionary reading in minute analyses of style. His critiques of Wagner's Hegelisms and other philosophemes provide the most interesting examples of this sort of reading; these work at once a cross-fertilization and a mutual destruction of art and philosophy.

Nietzsche's emphasis is always on turning, on keeping meaning and style in motion. Thus, while he sometimes describes his own relation to Wagner as an *Überwindung,* he celebrates the devalued and virtually invisible aspects of Wagner's revolutionary style in a manner that frustrates any final judgment, any clear dissociation of artist from critic. His reading of the operas and of Wagner's criticism makes Wagner's "great style" seem ridiculous for not supporting its great burden of propaganda. Yet, ironically, therein lie Wagner's greatness and his lasting contribution to music and all art. The operas cannot enthrall music to philosophy; according to Nietzsche, they prove that "art as idea" is an untenable hybrid and, more important, that art and language cannot spring free of philosophical Idea. Of course, this is not to say that Nietzsche deluded himself with the formalist's belief that art could exist *pour l'art.* Nietzsche did not simply devalue Wagner's place in philosophy; rather he reexamined—indeed "transvalued"—the position of modern art vis-à-vis its philosophical inheritance.

Nietzsche's overcoming or transvaluation of Wagner came neither easily nor all at once. Throughout his career, he fought equally against simply rejecting and wholeheartedly accepting the man he always acknowledged to be the greatest living artist. Wagner was the subject of the largest part of his aesthetic—and autobiographical—writings (even though Heidegger seems intentionally to diminish this fact when formalizing Nietzsche's aesthetics). Nietzsche's aesthetic writings, violently antisystematic as they sometimes are, depart from the margins of Wagner's writings to comprise layers of revision and self-critique. He continued appending "postscripts" and "prefaces" to his own literally marginal texts that would never appear as formal, proper, completed books. Some of his most challenging writings are virtually marginalia to modern art and philosophy, especially Wagner's and his own.

The obvious example of Nietzsche's habitual second-guessing and double-binding is *The Birth of Tragedy,* his first and most internally coherent book. We are thrown off course before we begin reading even this relatively manageable volume by "An Attempt at Self-Criticism," the Preface Nietzsche added fourteen years later to the second edition of his most influential and best-selling text. This preface is devoted almost exclusively to Wagner, direct consideration of whom had originally occupied only a brief panegyric at the end of the argument, as a sort of application and hopeful sign of a practical (re-)birth of (a theory of) tragedy. Perhaps more noteworthy than that belated preface's emphasis on Wagner—an emphasis that became more important if more oblique as Nietzsche's career progressed—is its title. "Self-Criticism" suggests the close if troubled intimacy Nietzsche felt with Wagner. Indeed, as we shall see, this intimacy of critic and artist was strategically explored by Nietzsche, who hoped thereby to make philosophy more poetic and to free art from Wagnerian (which is to say, as he does, "Hegelian" and "Schopenhauerian") ideology. *The Birth of Tragedy*'s status as rhetorical, ironic, polemic, and tropological text has received a good deal of critical attention by de Man and a host of Europeans before him. But we can also find its questions, and begin the questionable reading procedure it provokes, in Nietzsche's more *rigorous* (a word which, by now, should also suggest de Man's supplementary adjectives, *errant* and *erroneous*) later writings on Wagner.

The *Case of Wagner* is full of contradictions and layered by afterthoughts, apologies, and explanations loosely gathered into its two postscripts, an epilogue, and a preface that with apparent (self-)irony insistently treat Wagner dismissively. Instead of correcting his analysis, these substitutions add complications which carry it across the borders of several disciplines to confuse the genre of Nietzsche's own text. It should be recalled—Nietzsche never allows one to forget—that "*der Fall*" means "failure," "ruin," "downfall," the English and theological "fall/Fall" as well as "condition" and the medical or legal "case." The whole work, while intended to expose Wagner's secret failure or "lie," turns on the metaphor of a medical diagnosis. One might say that its metaphors literalize a physiology against the essentializing of art. For Nietzsche, Wagner is simply the most interesting victim of the common modern "disease" of nihilism. He can only hope for an inoculation against the naysaying moralism and decadence behind Wagner's dream of fabricating a synthetic art from fragments of (German) culture and philosophy. Nietzsche's most effective strategy is parody, a gay science (*Frölische Wissenshaft*), as antidote for Wagner's high seriousness:

> I have granted myself some small relief. It is not merely pure malice when I praise Bizet in this essay at the expense of Wagner. *Interspersed with many jokes, I bring up a matter that is no joke. To turn my back on Wagner was for me no joke. To turn my back on Wagner was for me a fate; to like anything at all after that, a triumph.* Perhaps no one was more dangerously attached to—grown together with—Wagnerizing; nobody tried harder to resist it; nobody was happier to be rid of it. A long story! *You want a word*

for it?—If I were a moralist, who knows what I might call it? *Perhaps selfovercoming.—But the philosopher has no love of moralists. Neither does he love pretty words.* (CW, 155; emphasis added)

On the ruins of Wagnerian moralism, that aspect of Wagner and his cult he most abhorred, Nietzsche hoped to build his affirmative but not necessarily systematic philosophy. His notion of system building—of philosophical constructs as well as architectural metaphors—is clearly different from those of other philosophers.[22] Nietzsche does not demolish Wagner's system to erect his own; rather, he finds that he can question System generally by anatomizing and then borrowing the fragments that unsettle Wagner's colossal works. More rigorously even than Heidegger, who still wants to build his very own system on the solid ground (*Grund*) of concepts, Nietzsche employs and ironically privileges those figures which disturb philosophy and philosophical art at their very foundation. It should be noted that, in the above passage and elsewhere, he subtly parodies the moralist's stance by using the word "*Selbstüberwindung,*" a self-conquest which in his critique is a denial of identity itself, thereby blurring the distinction between reader and writer or critic and artist. Obviously, he does not conquer himself or Wagner in this essay. Instead, he turns, twists, parodies, and otherwise ironizes Wagner's style and his own earlier idealized readings, which can nevertheless still be detected in this revisionary essay. One must attend carefully to minutiae like Nietzsche's choice of ambiguous and paradoxical words. The success of his critique depends upon such details, which mark the complex history of his opinions by recurrent metaphors/concepts/ideas, like "overcoming" and "style."

Before examining the formal details of the *Case of Wagner* or citing Nietzsche's formalization of the "question of style," it is helpful to look briefly at the twisted and discontinuous history (or autobiography) of his own Wagnerianism. The moment of his break with Wagner cannot be precisely located; on the other hand, neither can his moment of uncritical embrace or idealization. Nietzsche's self-reflections undo the clear distinction between art and interpretation, frustrating even our modest hopes of framing Nietzsche's own theory of art in a history of disillusionment. His own rereading (and/or rewriting) of the first full-length analysis of Wagner, *Richard Wagner at Bayreuth: The Fourth Untimely Meditation* (1876), has wider implications for the development of his readings in the history of philosophy:

> Instinctively *I had to transpose and transform everything into the new spirit that I carried in me.* . . . . In all psychologically decisive places I alone

am discussed—and one need not hesitate to *put down my name or the word "Zarathustra" where the text has the word "Wagner."* The entire picture of *the dithyrambic artist is a picture of the pre-existent poet of Zarathustra,* sketched with abysmal profundity and without touching even for a moment the Wagnerian reality. (EH, 274; emphasis added)

Confessing to be a more than unreliable narrator, a too subjective critic, he confuses the roles of reader and writer, as well as the identities of "Wagner" and "Nietzsche," in the fictional character or persona Zarathustra, who is the "new" and mature "Nietzsche" behind that poet-philosopher. This brief portrait or "lyric moment" (such reflections, though inappropriate to philosophy, are common enough in certain types of modern poetry) economically indicates Nietzsche's transformative relation to Wagner. It also describes the irregular rhythm of all interpretive transactions (literally a sort of dithyramb).[23] There is more to Nietzsche's psychological and aesthetic studies of Wagner than such tantalizing (self-)reflections back onto the reading act. He also makes direct programmatic statements about interpreting first Wagner and then all art and philosophy.

From the earliest eulogies to his (never quite) total renunciation of Wagner's "lie of the great style" (CW, 157), questions of interpretation and language explicitly merge with his attempt to define his own effort at constructing a new philosophy. Undoubtedly, *Richard Wagner at Bayreuth* exemplifies Nietzsche's own (Romantic) nostalgic hope of recovering a more vital language by reversing the exhaustion that robs modern language of its emotional impact—not to say its adequacy to the things of nature. His early praise seems to take Wagner's operas at face value. He mistakes that artist for a serious critic of dead ideas; concerning the problems of audience and media, Nietzsche envisions the perfection of German culture in a distinctly Wagnerian mixture of primitivism and innovation. He uncritically reads a synthesis of the Wagnerian doctrines of *Gesamtkunstwerk* (totalized art work) and *Zukunftsmusik* (music of the future) in the "happenings" at Bayreuth. He endorses Wagner's proposed marriage of drama, poetry, music, and the visual arts in new (*original*) works that might somehow return to encompass the perfection of an *original unity*. At this phase of his critical development, Nietzsche seems content with the circularity of his own hermeneutic and with Wagner's ability to reflect the desires of his audience in another nearly perfect (hermeneutic) *circle*. Because we read back from our own perspective and therefore through Nietzsche's later works and their subsequent interpretations, we sense the potential for an ironic treatment of Wagner in the following passage:

> For an event to be great, two things must be united—the lofty sentiment of those who accomplish it and the lofty sentiment of those who witness it. . . . *This reciprocity between an act and its reception* is always taken into account when anything great or small is to be accomplished; and he *who would give anything away must see to it that he find recipients who will do justice to the meaning of his gift.* (RW, 101–2; emphasis added)

While he nominates himself as privileged reader and teacher, the insider or privileged reader who narrows the distance between artist and audience, Nietzsche also opens a gap between art and interpretation, which admits all manner of play and discloses the specter of "Nietzschean reading." This din of rival interpretations and competing aesthetics might lie somewhere behind Pound's allusion to be the "neo-Nietzschean clatter," which the departing poet of "The Age Demanded" cannot integrate into "His sense of graduations" (Per, 201). Nietzsche's disruptions remained part of Pound's critical repertoire; even after World War I that German seemed untimely, not at all what the age demanded.

At the same time that Wagner, whom Nietzsche came to mock as too much the child of his time and nation, found adoring audiences at Bayreuth and at Ludwig II's fairytale castles, Nietzsche's ever more critical readings came to abuse meaning and to evade easy acceptance. Eventually he privileged disruptive moments in Wagner's—and his own—works precisely for inaccessibility and "dissonance" (both musically and more generally) from traditional taste. His reader's revenge against Wagner is strategic in compensating for Wagner's betrayal of art for political and economic expediency. Nietzsche says as much: "What did I never forgive Wagner? That he *condescended* to the Germans—that he became *reichs-deutsch*" (EH, 248).

Wagner's transformation from revolutionary artist to apologist for, even creator of, tradition is hardly unique among modern artists. T. S. Eliot's modern Anglo-American career is an exemplary case. Beginning as it did with a poem ("The Love Song of J. Alfred Prufrock") so revolutionary that Pound had to intimidate Harriet Monroe into printing it in her relatively progressive journal, *Poetry,* and ending with the Nobel Prize, Eliot's trajectory inscribes a pattern of growing conservatism that seems paradigmatic of what recent critics have termed "classical modernism." Pound's reading of "The Reverend Eliot" leads us to such a conclusion. One cannot make an easy equation or an exact parallel between the two German proto-modernists and the American modernists. The relationship between Pound and Eliot is no less complex than that between Wagner and Nietzsche, and the complexity is compounded by the fact that Pound did not limit his attacks on the modern tradition to his former revolutionary cohort. Instead, Pound searched through sinology and French letters and American politics and beyond in order to attack a literary tradition of which his old friend Eliot was only the most important, representative critic.

Before Pound, and in a more singular and unrelenting fashion, Nietzsche debates the virtues of Wagner's efforts to find an accessible style in terms of the artist's willingness to reflect the audience's desires. Defending his own refusal to achieve a popular style, he audaciously raises his own criticism to a higher status than Wagner's art. He never simply lets Wagner alone; his most successful critiques take the form of subtle parodies and thereby depend on or even incorporate Wagner's themes and style. For example, in order to expose Wagner's sexual repression and the myths of Germanic purity, he tropes himself into the figure of Siegfried, the virgin knight, who appeared parentless from an unknown past. In *Ecce Homo,* Nietzsche adopts the counter-persona, or the transformed persona, of a "premature birth of a yet unproven future" (EH, 298).[24] Elsewhere, against Wagner's clear messages and promises, he adopts a coyness that continually frustrates the reader's desire to find univocal readings behind his difficult style and surprising interpretations:

> One does not only wish to be understood when one writes; one wishes just as surely *not* to be understood. . . . All the more subtle laws of any style have their origin at this point: they at the same time keep away, create a distance, forbid "entrance." (GS, 343)

That might almost be read as a definition of Nietzsche's method of discovering the hidden Wagner—because, he suggests, for all that musician's overstatement, Wagner's best work was distant and so subtle that Wagner himself could not enter it. As we shall see, this aesthetic distancing (a word, especially with a pun on the neologism, *dis-tanz*[*ing*], favored by Nietzsche-Heidegger-Derrida) becomes Nietzsche's rhetorical trope against, or dance away from, Wagner's "oratorical art." Even before he came directly to criticize Wagner's use of art as rhetoric or propaganda, Nietzsche first privileged his strategies, at once adopting and adapting them.

For instance, in an identifiably Romantic mode (and here the Derridian supplementary adjectives "guilty" and "Rousseauistic" are clearly in order), Nietzsche rejects philosophical abstractions for the ideal of a poetic language (both poetry and music) that might recapture the passion and expressiveness of primitive speech. At the same time, his own writings become increasingly metaphoric, using figures to interrogate Wagner's notion of figuration. Incidentally, the Symbolists adopted this Rous-

seauistic-Schopenhauerian-Nietzschean-Wagnerian notion of language.[25]
Perhaps by indirection, a certain troubled nostalgia for a primordially musical poetics might be seen trickling down through Remy de Gourmont
and into Pound's dream of recuperating Provençal lyric poetry—though
Pound emphasized the interpretative over the primitive, even in *The Spirit
of Romance*. Again, Nietzsche represents a fuller range of acceptance and
rejection of the mythic and primitive than does Pound. The following
passage exhibits clear traces of the expressive and symbolic theories of
language associated with various Romantic writers. Yet Nietzsche's own
metaphorics of a physiology of art can be seen to adumbrate his later
demythologization of Wagnerian Romanticism:

> He [Wagner] was the first to recognise an *evil which is as widespread as
> civilization itself* among men; *language is everywhere diseased,* and the
> burden of this terrible disease weighs heavily upon the whole of man's
> development. Inasmuch as *language has retreated ever more from its true
> province—the expression of strong feelings,* which it was once able to con-
> vey in all their simplicity. (RW, 132–33; emphasis added)

In view of his subsequent interrogations of primitivism and onto-theo-
logical Truth, such talk of evil and the lapse from primordial simplicity
startles. Still, Nietzsche's mature critiques of Wagner and the theory of
language we find there are not simple reversals of these early enthusiasms. He never completely rejects these old values in favor of either a
revised nostalgia or a more revolutionary modernism. Vestiges of Wagner's grandiose schemes remain at least implicit in Nietzsche's analyses
of the conflicting desires for continuity and innovation within all Western
philosophy. A rigorous questioning of tradition and the status of art emerges
from his apparently erroneous preference for the lacunae in Wagner's
artistic system and musical oeuvre.

Indeed, Nietzsche founds his definition of modernism on this "contradiction of values," on the dissonance of kitschy details and grand ideas.
For example, he ironically characterizes Wagner as the representative,
even the "Redeemer" (a troping of Christ, as "falsehood" "made flesh"
rather than the Word, Logos, or Truth made flesh), of modern man. At
the same time, he equates the redeemer with Cagliostro, an eighteenth-
century charlatan most noted for parlor tricks, con-artistry, and speaking
in tongues:

> *Biologically, modern man represents a contradiction of values;* he sits be-
> tween two chairs, he says Yes and No in the same breath. Is it any wonder
> that precisely in our times *falsehood itself has become flesh and even ge-
> nius?* [emphasis added] that *Wagner* "dwelled among us"? It was not with-
> out reason that I called Wagner the Cagliostro of modernity. (CW, 192)

Nietzsche discloses unacknowledged nihilism, a negativity so deceptive that it was the motive force behind the spectacular operas. Unlike the nihilism charged against Nietzsche's destructive readings of philosophy, Wagner's "No" affirms traditional theology and the mutually supportive relationship of art and metaphysics. Wagner says no to the unaccountability of music and the freedom of the artist to explore new forms; instead, he simply fulfills the culturally and politically mandated expectations.

Nevertheless, as Nietzsche's reading shows, rather than culminating the Western tradition, Wagner produces Baroque conglomerations of confused Christianity and repressed sexuality. Nietzsche demonstrates that though Wagner had begun *The Ring* as an "immoralist," in the end the composer redeems his heroes from their violent, sensuous, mythical, and, in the case of Siegfried, incestuous origins. Curiously, Wagner betrayed art by ignoring the critical potential of his own early work. Still, and Nietzsche comes to value this failure, Wagnerian opera shows traces of a more revolutionary style that will not submit to bourgeois taste and the "modern" reinvention of theology in art.[26] Nietzsche does not wish to reverse Wagner's drift away from "pure art" but to question the enslavement of his art to old ideas. For Nietzsche, modernism marks an embrace of self-reflexion and analysis in art, not a return to innocence or the primitive. In Wagner he finds both versions of the new, and on the analysis and transvaluation of these two tendencies he founds his reading procedure.

Wagner is of inestimable value to the philosopher who would analyze modern contradictions—and do so with an eye to more than their easy resolution into a higher Truth or, in Nietzsche's terms, that new nihilism behind the reinscription of Christianity in *Kultur*. Rather than posit a Truth against Wagner's "falsehood," he reads through moral posturings to reveal and reprivilege the heterogeneous fragments of "classical art" and "noble" morality that argue against the illusory synthesis of art with German philosophy and Christian morality.[27] If he interpolates affirmations of his own position that might not be apparent (or even present) in Wagner's works, Nietzsche's arguments are all the more potent. Underscoring the disruptive and barely legitimate aspects of the operas, biography, and criticism, Nietzsche gratefully employs Wagner as a tool against modern complacency and as the chief representative of the modern moral disease—morality itself. Nietzsche implicates himself and his readers in an undoing of the category as well as the privilege of "modernism":

> But all of us have, unconsciously, involuntarily in our bodies values, words, formulas, moralities of *opposite* descent—we are, physiologically considered, *false*. *A diagnosis of the modern soul*—where would it begin? With

a resolute incision into this instinctive contradiction, with the isolation of
its opposite values, with the vivisection of the *most instructive* case—the
case of Wagner is for the philosopher a windfall—this essay is inspired,
as you hear, by gratitude. (CW, 192)

Even if we choose to read these final words of his epilogue as a straightforward promise to "isolate opposite values," Nietzsche's reading works
by showing that opposing values inhabit each other to subvert all systems
and all apparently distinct and stable oppositions. As Nietzsche recognized, his own resistance to Wagner's ideas necessarily implies a dependence on the onto-theo-logical system that Wagner championed.

Wagner's strategic importance lies as much in his own unconscious
sub-version of his project (those stylistic and historical details under [sub]
the official version of his art), not simply in the seductiveness of an illusory totalization. While Nietzsche cannot dismiss the psychological and
rhetorical temptations of the operas, he can praise the accidental "masterpieces" which reveal Wagner's self-deceptive (non-)mastery: "His
character prefers large walls and audacious frescoes. It escapes him that
his spirit has a different taste and inclination—the opposite *perspective*"
(NCW, 663). Nietzsche's argument, a troping of Wagner's "vision" of
totalized Art, can be seen to encompass even more media and genres than
Wagner hoped to include in his works. These perspectives are transformative; their transformations ("transvaluations") yield varying Wagners,
turning even ruins and "collapsed houses" into unlikely stylistic triumphs.[28]
Nietzsche's highly figurative or rhetorical diagnosis continues with this
portrait of the decadent modern artist as the self-reflexive painter/archi-
tect of the music-drama:

> [Wagner] prefers to sit quietly in the nooks of collapsed houses: there,
> hidden, hidden from himself, he paints his real masterpieces, all of which
> are very short, often only one beat long—only then does he become wholly
> good, great, and perfect, perhaps there alone. (NCW, 663–64)

Here Nietzsche is not merely describing *Leitmotiv* as a combination of
brief sections of notes or images repeated until they become thematic.
His reading does not yield simple instruction in Wagner's musical rhetoric
or compositional megalomania. On the contrary, he presents an entirely
different aspect of Wagner's stylistic accomplishments. Hence Nietzsche
privileges Wagner's unintentional genius for self-analysis and the production of unaccommodated details. While arguing that the musician is
too philosophical, he turns Wagner's ideas against philosophy, all the
time forgetting Art as a pure category.

Theodor Adorno—who, one should recall, was both composer and mu-
sicologist, the musical consultant for Thomas Mann's *Doctor Faustus*—

provides a technical analysis of *Leitmotiv* that suggests how Nietzsche turned this Wagnerian invention against the composer:

> Beneath the thin veil of continuous progress *Wagner has fragmented the composition into allegorical leitmotivs juxtaposed like discrete objects.* These resist the claims both of totalizing musical form and the aesthetic claims of "symbolism," in short, the entire tradition of German idealism. Even though *Wagner's music is thoroughly perfected as style, this style is not a system in the sense of being a logically consistent totality,* an immanent ordering of parts and whole.[29] (emphasis added)

Despite all efforts to achieve a musical and discursive totality through the repetition and development of themes transcending musical style, Wagner's leitmotifs work as readings or commentary on the mythic past he hoped to evoke, if not more generally on the very project of transforming music into Idea. Thus Nietzsche seeks out places in the operas where Wagner fails of philosophical reflection but succeeds in self-reflexion, turning him into an aliented "modernist" in a sense more familiar to twentieth-century poetry. This manner of reading a system against itself, of uncovering the unaccountable in accepted or canonical art, so fundamental to Nietzsche's criticism, adumbrates Pound's rooting around in what had seemed the established whole of Classical and Renaissance poetry. For example, Pound's Troubadours marked individual inventions, not punctuations of a classical ideal of lyric.

Nietzsche's incongruous or ironic Wagner is "our greatest miniaturist in music" (CW, 171). Still, Nietzsche is not content to take Wagner's works apart at their seams (and "seems" or appearances); he ironically and critically reprivileges and reinscribes those figures and accidents of language and style that make impossible Wagner's dream of a totalized, integrated, organic, and metaphysical art.[30] He does so in a way that forces one to recognize all art and philosophy as the desire for a system that might provide stable meanings and a clear message, things made impossible by the inherent dissonance of language and the play of substitution on which Nietzsche's own writings thrive.

Nietzsche's assault on Wagner and on the "Wagnerian taste" for philosophical symbolism disturbs the hierarchy that privileges Idea over style. Perhaps his most decisive blow against Wagner's total art is his denunciation of the operas as mere oratory, or an overblown rhetoric that fails to disguise Wagner's programmatic intent of purveying Hegel's "Idea" and the notion of Germany's world historical mission. Again, art and philosophy are not in a simple dialectical opposition—that would be too Hegelian for Nietzsche. Even though the following passage could be used to support a notion of *l'art pour l'art* as well as nostalgia for Classical simplicity, Nietzsche does not replace Wagner's dogmatism with its mirror image, an equally dogmatic formalism. Instead, he subverts both philosophy's (read "Hegel's") hope of using art as a potentially transparent medium for meaning and (modern) art's hope of escaping such exploitation and complicity. He accomplishes this double exposure by focusing on the "means" Wagner had to employ to convey his borrowed "meaning." Nietzsche offers this reading of Wagner's ideas:

> [Wagner] repeated a single proposition all his life long; that his music did not mean mere music. But more. But infinitely more.—*"Not mere music"* [emphasis in text]—no musician would say that. . . . He remained an orator even as a musician—he therefore had to move his "it means" into the foreground as a matter of principle. "Music is always a mere means": that was his theory, that above all was the only *practice open to him. But no musician would think* that way [emphasis added]. . . . He was his life long the commentator of the "idea." (CW, 177)

We might note that the English pun on "it *means*" (*es bedeutet*) and a "mere means" (*nur ein Mittel*) was unavailable in Nietzsche's German. Therefore, he probably did not intend the pun, at least not for the Germans. Our reading might be said to produce accidentally the ironic increment of translation. As will be shown, Pound also exploited this beneficial abusiveness of translations (his term was "interpretative translation") characteristic of the Nietzschean reading style. Of course, it is just like Nietzsche to demolish a whole set of assumptions with a pun, in a strategy that both modern poetry and poststructuralist philosophy have found a-*mean*-able.

Following out this strategy, we see Wagner become a "means" by which Nietzsche can question the whole philosophical tradition—but never a mere means, because, however unwittingly, he lends to Nietzsche a cache of words, formulas, indeed a whole reading style. If this turn from meaning to an interrogation of mediation seems abrupt, we should remember what is at stake here. In the ostensible reordering of the relationship between modern art and philosophy, Wagner—with Nietzsche's *help*—sacrifices his practice to a theory it cannot sustain. Nietzsche does not hereby depreciate interpretative art. Wagner, in his view, rejected "mere music" without being aware of the impossibility of clearly distinguishing meaning from symbolic mediation. He rejected "mere music" for "mere idea" and thus adopted the traditional metaphysical idea of language and art as mere vehicles, neutral conveyors of meaning. We should recall Heidegger's (Aristotelian) idea of metaphor as something that philosophy might occasionally borrow for interpretation. Nietzsche does not long entertain illusions about a stable, nonfigurative, more than rhetorical language—

any more than he demands that Wagner stay within the proper bounds of music. Instead, his critique clearly exaggerates these apparent failures and difficulties, often by imaginative or poetic analyses that cut across several discourses.

Mixing the physiological and aesthetic categories of "taste," Nietzsche demonstrates that neither art nor philosophy remains untouched by the exchange of ideas that passes through metaphor. It is also noteworthy that Hegel, author of the System, is the precursor of Wagner's music, which therefore cannot convey pure ideas any more than it can claim a legitimate artistic ancestry. These thefts between systems might not be immoral, but, what is more embarrassing to Wagnerians, they confuse generic, cultural, and national distinctions to yield a mixed (but never a totalized) art:

> Let us keep morals out of this: Hegel is a *taste*.—And not merely a German but a European taste.—A taste Wagner comprehended—to which he felt equal—which he immortalized.—He merely applied it to music—he invented a style for himself charged with "infinite meaning"—he became the heir of Hegel.—Music as "idea." (CW, 178)

There Nietzsche would seem to subject Wagner to an exaggerated or unreserved Wagnerian reading, one replete with a pan-Germanic genealogy and the imperialistic claims of "infinite meaning" (*unendliche Bedeutung*), a Hegelism that Nietzsche unleashes into infinite interpretation and interrogation. Again and again he exposes the vulgar or commonplace philosophemes that punctuate Wagner's work and motivate the operas' obvious symbols (*die grössen Symbole*) such as the "unconscious Spirit of the People" (*"der unbewüsste Geist der Volks"*). Insistently, he brings these assumptions to light, affecting both the art that carries the ideas and the ideas themselves. Of course, after this procedure, there remains no hope of rendering an idea-in-itself (*Ideen-an-sich*), let alone a thing-in-itself (*Ding-an-sich*) or, in Nietzsche's phrase, beauty-in-itself (*Schönes-an-sich*).

Reading through Wagner's symbols and his fragmented *leitmotifs,* Nietzsche nevertheless emphasizes the media over the ostensible message they convey—which is not to say that he dismisses Wagner's bombast without irony. Nietzsche's Wagner is an ideologist whose ideas outrun or exceed their metaphysical, not to say jingoist, function. He transforms Wagnerianism into an assaultive self-parody of serious philosophy. Thereby he not only accepts Wagner's invitation to read art (like philosophy) for meaning; he also uses elements of that same art to interrogate both artistic media and the meaning-giving act.

Nietzsche does not refute Wagner's "lie of the great style" (CW, 157), though he exposes it as a fabrication, "patchwork, 'motifs,' gestures, formulas" (CW, 177). Rather than mourn the lapse from some ideal art, Nietzsche discovers his own distinctive style in the fragments—the symbols, motifs, metaphors, and allusions—that erupt beneath the surface of the massive Bayreuth spectacles to prove that music and the other arts cannot be subject to or mastered by philosophy—least of all by a metaphysics conceived along the old discursive lines.

As has been shown, Nietzsche's many books and autobiographical reflections on Wagnerian modernism do not satisfy the requirements of a systematic aesthetics. Yet his writings are more controlled, more self-conscious, and, notwithstanding the efforts of several editors, more available than Pound's fragmentary critical writings.[31] Economically, if ironically, Nietzsche formulates a notion of modernism that seems to adumbrate Pound's strange mixture of modernism and nostalgia. But it is against the background of Nietzsche's well-documented relationship with Wagner, out of which grew his own style, at once an imitation and an ideology of fragments against any notion of totalization, that we must read the wider implications of what Nietzsche terms "merely . . . the question of style."

The irony of the following passage culminates in "merely," for style is the focal point of the Nietzschean critique. This is also the case with Pound's various manifestoes, prospectuses, and programs that seem merely to address stylistic details. In such readings, style is a category which comes to include, or at least to diagnose, ideology. Nietzsche's brief statements in this regard have come to bear the weight of conflicting ideologies and thus to epitomize modernist (and postmodern) styles of fragmentation in prose narrative, poetry, painting, music, and even interpretation. One must realize, then, that Nietzsche does not merely lament the loss of Presence, which he exposes as an illusory marriage of such apparent opposites as style and Idea. Nor can he be said to substitute for this nostalgia a gleeful insistence on stylistic play for its own sake. He does not, in other words, offer easy answers; instead, he opens the question that was to engage Pound as it still engages us. He says:

> For the present *I merely dwell on the question of style.*—What is the sign of every *literary decadence? That life no longer dwells in the whole*. The word becomes sovereign and leaps out of the sentence, the sentence reaches out and obscures the meaning of the page, the page gains life at the expense of the whole . . . the whole no longer lives at all: it is composite, calculated, artificial, and artifact. (CW, 170)

That describes Nietzsche's readings as well as Wagner's writings, musical and otherwise. Nietzsche uses his own modernism as a critical or textual strategy for interpretations which uncover embarrassing details and artifacts in the metaphysical tradition. Thus he affirms (and reinscribes) the very decadence and heterogeneity he seeems so clearly to have decried.

IV

To whatever degree Pound may be said deliberately to adopt either the Nietzschean critique or the Wagnerian recuperation (*mythos*) of totalized culture, he has implicated modernist poetry in that unresolvable dialogue of philosophy (or criticism) and art. As the following chapters will show, Pound employs what must be called Nietzschean (or "neo-Nietzschean") strategies to inscribe a Kulchur that is more a series of recurrent revolutions than the recuperated continuum that Eliot called tradition. Pound's aesthetic writings are not nearly as focused as Nietzsche's. While he too tended to move from recuperative to disseminative readings, from a totalizing to an analytic art, he did not have one contemporary with or against whom he worked out his readings—not even Gourmont or Eliot or Williams or Lewis, each of whom critics have chosen as *the* major influence on Pound.

More frequently, indeed more rigorously than current academic readings would lead one to believe, Pound poses questions of style in terms of tradition and fragmentation. But, and this is what some scholars have forgotten, he does not simply or nostalgically oppose a fragmented present to an ideal (of) tradition. Instead, he presents a tradition and, one might say with reservations, a "reading method" for creating tradition(s) which consists of fragments that seem deliberately to defy system, continuity, and totalization. Nor was he unaware of the disruptive potential and far-reaching implications of his reading and writing styles. As I have suggested with regard to the compendium of programs in *Guide to Kulchur,* he allows his text to be thrown off course by radical questions. To repeat, he frequently returns to the fundamental and uninterrogated assumptions of literary history: "What is literature?" (ABCR, 28), "Why Books?" ("How to Read," LE, 20), "What is language?" (ABCR, 28). Perhaps readers have been thrown off course by his habit of submerging such questions under the more popular stylistic arguments surrounding Imagism, Vorticism, and "the new." In any case, to such questions he offers many partial, conflicting, and highly metaphorical, which is to say unmethodical and antisystematic, answers.[32]

Deploying the most sophisticated methods of literary criticism and textual scholarship, Pound's best critics have found straightforward answers to these questions; and they have turned his fragmentary notes into books and theories which seem, however artificially, to center Pound in a "modern tradition" or his own "era" compatible both with traditional artistic principles and scholarly practice. While I do not mean to suggest a conspiracy of omission or misreading, many of the letters and other prose texts that argue against the legitimate, recuperative Pound lie literally unread in archives. Therefore, before turning to these forgotten Poundian texts, one might recall how this other Pound has been read into the academic literary tradition. Inevitably with reflection on Pound's notion of reading come questions of how he has been *read*—in the traditional sense of reading as the mastering of writing, the halting of the play of textuality.[33] Entertaining such questions is merely to give Pound's own interrogations their due—even if such a procedure might seem occasionally to approach the violence of transvaluation.

Here Hugh Kenner's enterprise is crucial. More than any others, his readings have mastered Pound's texts to make them coherent, if not academically acceptable. Indeed, his books are so persuasive that many readers use them as guides through, or even substitutes for, the complexities of Pound's poetry and prose. From these influential readings emerges a Pound who is the best exponent of Eliot's notion and essay "Tradition and the Individual Talent," a Pound who seems to have passed by means of a radical questioning into the mysterious if tenuous balance of tradition and innovation Eliot articulated. One must admire how carefully Kenner reads Pound and how skillfully he substitutes for the more subversive aspects of Kulchur Eliot's still potent organic figuration of "classical modernism" as the culmination of a "simultaneous order" stretching from Homer to the present. Still, in all its guises and perhaps largely because of its metamorphic character and figurative names, Pound's Kulchur is kinematic, more a disturbance in translation than a series of perfectly translatable monuments.

For Kenner's clearest statement of Pound's virtually Eliotic (even neo-Aristotelian or Aquinian) method and intentions, one has recourse to his first, indeed *the* first, full-length academic study of Pound, *The Poetry of Ezra Pound*.[34] Kenner continues to elaborate its thesis through Pound's exemplary texts, and the respectability that book won for Pound was the basis of the authority that would later position the poet centrally in Kenner's history of modernism, *The Pound Era*.[35] I would not depreciate

Kenner's contributions to, his virtual founding of, Pound studies. In fact, one can no more break free of Kenner's basic methodology than Derrida can free himself from (Heideggerian) readings of Nietzsche which would make that philosopher part of the metaphysical tradition he critiqued. Indeed, in view of the intractability of Pound's contradictions, Kenner's reading must remain a temptation. Who would not welcome a more reasonable, formalist, apolitical, or at least civilized Pound? Just as one must submit Nietzsche to careful systematization before his aberrations appear fruitful, one must bear with traditional readings until Pound's own violations of such methods point the way to a different, but not wholly oppositional, reading procedure.

As prelude or *pre-text* to opening a rigorous interrogation of Pound's reading program, we should recall Kenner's sleight-of-hand or ventriloquism in substituting Eliot's "tradition" for Pound's more radical notions of reading.[36] Eliot's formulation is too familiar to require quotation, but this ascription of it to Pound warrants consideration—if only for its telling onto-theo-logical privileging of voice and originary poetry one has learned to read as a version of mythopoesis. Nevertheless, Kenner at once dodges and underscores the problematics of a Nietzschean (which is to say, following Pound, a disseminating and palimpsestic) reading. Crossing poetry and prose, cross-referencing Pound and Eliot, Kenner is forced to accept a subversive notion of influence when he makes Pound speak Eliot's words:

> There should by now be no difficulty in seeing how Pound, who introduced the author of *Little Gidding* to the permanent significance of Dante, cannot be neglected as an implicit component in the "familiar compound ghost" who among the ruins of a burning world confronts the fire-watching Eliot in incarnate magistracy: Tradition confronting the Individual Talent.
> Since our concern was speech, and speech impelled us
> To purify the dialect of the tribe
> And urge the mind to aftersight and foresight. . . .[37]

CHAPTER TWO

The Will to Read
Pound's Revolution in Pedagogy

America should have a great interpretive literature, and . . . men should have the opportunity of making experiments conducive to that literature.

POUND, 1915[1]

In the year 1930 we are or shd be faced with the perception that literature has fallen behind science. . . . And this crisis shd carry such of us as are readers and certainly such as are writers to a closer consideration of language.

POUND, 1930[2]

In his search for an "interpretive" poetry or poetic that would effect "a revolution in the study of literature" (LE, 19) by linking those antithetical categories of the critical and creative, Pound makes a decisive "turn toward language." Despite his invocation of a scientific model, of what has seemed to some a simple empiricism and an "objective" poetics (one interpretation of Imagism), he insistently poses questions of reading and writing in terms of "the nature of language." Both "How to Read" and *ABC of Reading* depart from and return to questions from which most literary critics have retreated: "What is *literature, what is language, etc.??*" (ABCR, 28).[3]

Though Pound does not retreat from these questions, he confronts them in the only possible way—indirectly, that is, metaphorically. Pound's various critical answers to those deceptively simple questions recur as metaphors that call attention to their own metaphoricity, thus complicating the easy distinction between poetry and criticism that literary history has always assumed. Despite the popular view that Pound's theory of language, or of poetic language, is organic and argues for a natural origin of language, he repeatedly asserts the opposite, that language is instrumental and belongs as much to culture (and science) as to nature: "LANGUAGE was obviously created, and is, obviously, USED for communication" (ABCR, 29). Pound's notion of language, however, does not involve the communication of a stable meaning, carried intact from one point or person to another. Instead, through a series of metaphors drawn from the "new sciences" of electromagnetism and telecommunications, he equates (but in an unstable or differential equation) language with "force," "movement," and "power"—terms familiar from the Nietzschean lexicon. The source and terminus of this power are ambiguous, even interchangeable. Pound further augments this potential for transformation and distortion, giving it the privileged name "literature": "Literature is *language charged with meaning*. 'Great literature is simply *language charged with meaning to the utmost possible degree* [E.P. in *How to Read*]'" (ABCR, 28; emphasis added).

Literature partakes of the movement or tropology of all language. To read poetry is to engage in an exchange of power, of interpretations "charged" or activated by reader and writer, but also by the language they share. For Pound, languages are transformational and plural; since every reading involves a translation, none is capable of univocity or totalization: "The sum total of human wisdom is not contained in any one language, and no single language is CAPABLE of expressing all forms and degrees of human comprehension" (ABCR, 34). Reading is never a simple matter of facts, of conveying discoveries and correcting interpretations. Instead, it is a matter of will and the exertion of power; in Nietzsche's sense, the "Will to Power": "Facts is precisely what there is not, only interpretations. We cannot establish any fact 'in itself': perhaps it is folly to want to do such a thing. . . . In so far as the word 'knowledge' has any meaning, the world is knowable, but it is *interpretable* otherwise, it has no meaning behind it, but countless meanings—'Perspectivism'" (WP, 267). While Pound stops short of that perspectivism, he too recognizes an exchange of power at the scene of reading: "we shd read for power. Man reading shd be man intensely alive. The book shd be a ball of light in one's hand" (GK, 55). The reader comes to books "for power," yet the books, Pound insists, which one carries and presumably has already read, illuminate or interpret the books one will read. Faced with this textual play, Pound hopes not to abolish the canon and canonical readings, but to reread, even to interfile them, thus enhancing their power and luminosity: "the great classics inter-illuminate one another" (SP, 30).[4]

Pound's reading procedure is transformational or, better, translative. It offers no escape from the intertextuality of reading to a "thing" or a "truth" before language and interpretation. His claim that translations from one language to another, from one epoch or genre to another, are interpretations, not the transfer of fixed meaning through transparent media, leads him to equate reading, translation, metaphor, and even criticism. He alludes to Dante's translation of the classics as a form of criticism thus: "Dante who was capable of executing the work and of holding general ideas, set down a partial record of procedures. 2. Criticism by translation. 3. Criticism by exercise in the style of a given period" (LE, 74). Pound's own reading textbooks advocate both critical methods. In order to achieve the marriage of theory and practice he attributes to Dante, he proposes classroom exercises in translating traditional poetry into modern culture. Moreover, as early as *The Spirit of Romance*, "art" and "interpretation" and "metaphor" are inextricably interwoven in Pound's critical lexicon: "An art is vital only so long as it is interpretative. . . . The interpretive function is the highest honor of the arts. . . . *The Divina Commedia* is a single elaborated metaphor of life, it is an accumulation of fine discriminations arranged in orderly sequence" (SR, 87–88). When he refers thus to reading and/or writing as comprised of "interpretative metaphors," Pound necessarily affirms that as metaphors they too require further interpretation—or reading.

In this way, "How to Read" apparently loses sight of its titular promise of a reading method or curriculum when Pound substitutes "writing" for "reading" in order to underscore his aim of "charging" language with power: "Let us try to keep our minds on the problem we started with, i.e., the art of writing, the art of 'charging' language with meaning" (LE, 29). Again and again, which is not to say without revision, he will convey his notion of reading in metaphors that undermine the philosophical or scientific methods he otherwise advocates, methods which have always mistrusted language and the metaphorics Pound emphasizes: "This book can't be the whole history. Specifically we are considering the development of language as a means of registration. . . . The present booklet is concerned with language" (ABCR, 56–57). The focus on language does not, in fact, deflect interest from the topics of reading and literature. Pound deliberately opens his text to slippages among various discourses and languages by denying that he is a philologist (his derisive word is "philologer"), a linguist, or a historian. He insists instead on treating literature in its own terms. Refusing what has always been the first step toward method, he rejects the possibility of a meta-language and a criticism fully divorced from the texts under analysis. After distinguishing his procedures from morphology and the reading of "laundry bills," he says that his own "method has nothing to do with those allegedly scientific methods which approach literature as if it were something *not literature*, or with scientists' attempts to sub-divide the elements in literature according to some non-literary categoric division" (LE, 19).

More than one critic has felt tempted to read such programmatic statements as endorsements of the old New Criticism that treats poetry as a thing-in-itself or, as Eliot would say, "autotelic." Pound's method is something quite different, and, by virtue of his notion of language as a generative force, it is uncontainable and unmethodical even by the standards of poetic form or formalisms. The disruptive and disseminative force of language becomes most apparent in repeated gestures toward method that end, not in system, but in the proliferation of borrowed readings and procedures. For example, here Pound at once confuses and distinguishes scientific and poetic methods by introducing the question of a language that substitutes graphs for speech and Ideas, thereby necessitating interpretation rather than reading: "Fenollosa emphasizes the method of science, 'which is the method of poetry,' as distinct from that of 'philosophic discussion,' and it is the way the Chinese go about in their ideograph or abbreviated picture writing" (ABCR, 20). Pound recognizes that methods, like languages, are multiple and transformational but not equivalent; that is, they do not translate a stable meaning from one text or discourse to another but in effect "charge" or supplement discourses. Methods extend and displace rather than contain texts. Reading generates but is not genetic; what it organizes it cannot master.

However programmatic, statements like those just quoted resist any program, including the one exemplified by Hugh Kenner's argument that Pound envisions a literary tradition or canon as a complete system of differences related by "nodes," "subject rhymes," or Buckminster Fuller's "knots," which are a "self-interfering" yet accumulative and organically developing structure of classics pointing toward "accuracies" or truth. Side-stepping Pound's most radical statements that languages are irreducibly differential and metaphoric, *The Pound Era* affiliates the poet's "theory" of language with nineteenth-century German philology and, perhaps not accidentally, with the genealogical model of etymology—which might be thought to resemble historical interpretations that establish literary sources and canons.[5] Specifically, Kenner affirms the nearly mystical resolution (sublation) of the still independent languages into the whole "family" of a continuous and singular Indo-European language: "Words characterize languages, languages are discriminated phases of Language; Language is the total apprehension, in time and space of the human mind, the labyrinthine marvel. Philology, in sorting out such matters . . . permitted Pound's generation the vision of languages as inter-textured, cognate systems of apprehension, to each its special *virtu*" (PE, 120).[6] Kenner's assertion that Pound resolves the plurality of languages into a universal system, his narrative that ascribes a fixed and hermetic pattern to Pound's "intertextuality" and "intratextuality," seems to issue from a hope for totalized readings—and thus for the "totalitarian culture" that Pound himself held in reserve. Even Kenner, who has taken quite seriously Pound's notions of the "force" of reading and the "force field" of language, has retreated from the problematic of language elaborated in the reading texts. Therefore, we must (again and still) ask, "What is Pound's notion of reading?"

Unlike current "reader-response" critics, Pound does not grant a constitutive role to the reader, nor does he formulate a hermeneutic which might recover a writer's initial intuition, as does the Heideggerian or phenomenological model of the hermeneutical circle. Quite simply, his notion of language forbids either binary or dialectically resolved models. For Pound, "reader" and "writer" are, like "noun" and "verb," grammatical fictions or functions in a language that is radically transformational and translative. *The Chinese Written Character,* frequently invoked in *ABC of Reading,* affirms this "translational" (Pound also uses the adjectives "metamorphic" and "metaphoric") effect of all linguistic exchanges. For example, in a footnote to a passage calling upon modern translators to "feel back along the ancient lines of advance," the connections or organic lines by which poetic metaphors progressively unveil Nature, Pound (or is it Fenollosa? or both?) locates the complex reading-translation-metaphor in the most ancient texts, in Nature as well as modern Western languages:

> I would submit in all humility that this [advance backward through metaphors] applies in the rendering of ancient texts. The poet, in dealing with his own time, must also see to it that language does not petrify on his hands. He must prepare for new advances along the lines of true metaphor, that is *interpretive metaphor,* or *image,* as diametrically opposed to untrue, or ornamental metaphor. (CWC, 23; emphasis added)

It is at least noteworthy that Pound inserts "image"—if not *Imagisme*—into a discussion of metaphors and interpretation. This would tend to belie both his notorious abhorrence of "rhetoric" and his claims to an "objective" language.

Pound's attention to the rhetorical economy, to language as a "medium of exchange" between readers and writers, as contrasted with an emphasis on "reader" and "writer" as stable identities or mutually exclusive perspectives, has profound implications for theories and histories of reading. Though not with reference to Pound, these implications have been explored by such recent interpreters of Nietzsche as Paul de Man, who, in a subtle critique of Hans Robert Jauss, the founding practitioner of *Rezeptionsästhetic* (which attempts a "paradigmatic condensation within a diachrony" or, more simply, a history of changes in the reception of literature by stabilizing "horizons of expectation"), shows that Walter Benjamin's model of translation radically destabilizes the fixed meanings on which literary history and such aesthetics as Jauss's depend:

> By invoking the "translation" rather than the reception or even the reading of a work as the proper analogon for its understanding, the negativity inherent in the process is being recognized: we all know that translations can never succeed. . . . "translation" also directs, by implication, the attention to language, rather than perception, as the possible locus for this negative moment. For translation is, by definition, interlinguistic, not a relationship between a subject and an object, or a foreground and a background, but between one linguistic function and another.[7]

De Man here, as in other contexts, invokes a Nietzschean distinction in order to read Benjamin's notion of "translation" as a "theory" of criticism. Theory, he argues, only begins when one turns away from the historical, aesthetic, or psychological models of literature, toward language itself. If traditional criticism has been a "detour or flight from language" (AR, 79), then the return to language will not simply be an alternative model, but it will, as Benjamin's theory of translation suggests, uncover the intralinguistic problematic of a negative moment or *aporia* in the linguistic function. Literature, for de Man, repeatedly exposes the referential illusion.[8] In the face of literature's own problematics, linguistics or rhetoric cannot yield a "science of literature"; on the contrary, focus on the transformative, that is, the tropological as well as the translative, character of language confirms the instability of all models, including the scientific.

Pound's yoking of reading and language, his analyses and deployment of metaphors of translation and transportation among texts and within culture, fundamentally questions the traditional critical procedures that have been ascribed as well as applied to Pound's text. To the contrary, he entertains—if not systematically, at least repeatedly and from several perspectives—notions of language and rhetoric that explode from within the hierarchies and taxonomies he offers by way of reading guides and curricula. Poundians have not failed to remark (and remake) his statements about language, especially Imagism's alleged rejection of rhetoric for a direct, univocal, or "objective" poetry. But most critics have granted authority to Pound's lists and *method(s)* at the expense of the metaphors that undermine them. It is, then, with Pound's use of metaphor, and in particular his reading of Aristotle, that one must begin to "read" his reading texts—but not without a certain neo-Nietzschean indirection.

II

At a crucial point in the *ABC of Reading,* where Pound offers his desiderata for reading instructors, he alludes to one of Aristotle's definitions of metaphor: "Aristotle had something of this sort in mind when he wrote 'apt use of metaphor indicating a swift perception of relations'" (ABCR, 84). That reading, for it can hardly be called a citation or literal translation, of Aristotle is unusual—in fact, it sounds more Heraclitan than Aristotelian.[9] Aristotle's project was to identify and categorize tropes as structures of perception or cognition. But Pound seems to find in his definition more the activity of troping or turning, the force or process of a rhetoric that informs, than a perceptual field. For Pound, Aristotle represents a mode of action or process, not the stability of (Aristotelian) Categories. In other words, Pound emphasizes what de Man calls persuasive or performative rhetoric rather than rhetoric as a system of the tropes of cognition. By Pound's account, "perception of relations" is "swift," and it is the act, not the idea, that is "apt" for metaphor. Contrasted to accepted interpretations of Aristotle, Pound's reading underscores the differential or relational character of metaphors, specifically their speed of movement, the potential for anamorphic interpretation that requires "swift perception" by reader and writer alike. Pound ignores the hermeneutical or philosophical assumption that metaphors are either ornamental—that is, useful for interesting audiences in otherwise dull or inaccessible ideas—or instrumental, and hence used for conveying a stable meaning by means of two discrete components ("vehicle" and "tenor") which are resolved in readings that reproduce the writer's original "idea." Pound everywhere rejects the ornamental, though at times he does seem to argue for an "instrumental" notion of metaphor.

Pound also uses Aristotle's name to affirm the transformational qualities of metaphor he finds inscribed in, for example, the Chinese written character, ideogram or *ideograph,* which he values both as the best "medium for poetry" (as his subtitle for the Fenollosa/Pound essay suggests) and as an originary and even primordial graphic language. We should recall that Pound's ideograph (though one does not know where finally to draw the line between Fenollosa and Pound) opposes grammar to nature in such a way as to privilege rhetoric over logic and writing over speech.[10] Western grammar trammels nature by affixing names and functions to "action": "A true noun, an isolated thing does not exist in nature. Things are only the terminal points, or rather the meeting points, of actions, cross-sections cut through actions, snapshots. Neither can a pure verb, an abstract motion be possible in nature" (CWC, 10).

The passage that bears the footnote about Aristotle directs this grammatological analysis against Western poetics—even *The Poetics*—which defers to logic:

> You will ask, how could the Chinese have built up a great intellectual fabric from mere picture writing? To the ordinary Western mind, which believes that thought is concerned with logical categories and which rather condemns the faculty of direct imagination, this feat seems quite impossible. . . . *This process is metaphor,* the use of *material images to suggest immaterial relations.* (CWC, 22; emphasis added)

Pound's note transforms what might be an example of Classical rhetoric's notion of instrumental metaphors into a celebration of process and relativity, when he says: "Compare Aristotle's *Poetics:* 'Swift perception of relations, hallmark of genius.'" That definition, to be found nowhere in Aristotle, lays stress on movement and mediation (the mediation, for example, that Pound finds in the "time" of verbs as opposed to the space of nouns), focusing not on a specific usage but on the perceptions of the writer, the "genius." But the writer is also, in a sense, a "reader." For Pound, the two are never absolutely separable, nor are their "perceptions" reproducible as things seen. Still, we need to ask how Pound's theory of reading fits with Aristotle's definitions of metaphor. What can his reorientation of the founder of the philosophical Categories back into Chinese graphics mean, except an identification of the graphic with the metaphoric and metamorphic?

In effect, Pound has grossly abused Aristotle's authority in order to give language and Image priority over Idea and to privilege figural over literal meaning. Following Aristotle, philosophy has treated metaphor as if it were borrowed from rhetoric and/or poetry in order to make ideas clearer, or has condemned it for dimming the light of truth. Since language, especially as *figura,* is subordinate and temporally deferred to Idea and "object," metaphor or the rhetorical medium must be borrowed from poetry only to be discarded or effaced when it has served the purpose of clarifying ideas. Metaphor, according to this plan, is an unusual use of language, a momentary and reversible departure from literal meaning. *The Poetics* treats metaphor as "the application of an alien name," as metonymy. Denying any constitutive or trans-formational properties to language, Aristotle suggests the four ways in which metaphor can transfer unchanged *the meaning* "from genus to species or from species to genus, or from species to species, or by analogy, that is proportion" (Poetics, XXI.4). Pound's "swift relations" and "perceptions" are absent or repressed from the definition proper, but one detects them in the examples Aristotle uses for the types of metaphor. There, and in *The Art of Rhet-*

oric, the other major source of Aristotle's metaphorics, nearly all the examples of this transfer of meaning involve ships and transportation, specifically Homer's catalogues of ships. But these ships are not only the "best metaphors," they are also metaphors of metaphor, conveying their cargoes as metaphors carry meaning from word to word. Aristotle speaks metaphorically of metaphor, causing his definitions to rest on the play of signifiers, of which "metaphor" turns out to be but one possible and imperfect name. As we shall see, this same condition obtains in Pound's categorization of those writers who employ "interpretive metaphors."

Though Pound did not systematically treat the problems that inhere in rhetorical definitions which partake of the condition they describe, he was aware that Aristotle's imprecisions occasioned the commentaries of several Medieval rhetoricians (including Erigina, Aquinas, and Dante, all of whose criticism informs *The Cantos*), even though these failed to halt philosophy's continuing decline into imprecise and abstract language. Pound's condemnations of philosophical "jargon," sometimes under the name of "Arry," sometimes as "flyosophy," are commonplace. Of course, philosophy has never succeeded in purifying itself of metaphor or in taming language and imagination. Pound himself took advantage of this condition when he launched his Cantos on one of Homer's ships, by a translation (really a translation of a translation) of *The Odyssey* that begins in the middle of Homer's metaphors for metaphor and still has a distinctly non-Aristotelian application, especially since Pound's inaugural translation launches his poem out of another text. Translation tropes, turns, returns (to the underworld of subtexts) in order to go forward bearing the cargo of old texts transvalued. Aristotle notwithstanding, metaphor is translative and intertextual, and Pound had this from Homer.[11]

Pound's citation of Aristotle on metaphor has not escaped critical comment which has tended to affiliate *The Cantos* with organic and expressive theories of language, making his notions of reading and metaphor more properly neo-Aristotelian. For example, Hugh Kenner discovers Aristotle's *mimesis,* which requires stable meaning and transparent media of representation, and *praxis,* or human deeds and actions immanent in "plot," behind Pound's attempts to replace abstract with objective language and to argue for the superiority of the moving to the stationary Image. Thus, while he goes to those texts where Pound stunningly opens a problematic of language by his redefinitions of Aristotle, Kenner reinscribes Aristotle in Pound's poetics in order to complete—rather than interrogate—philosophical speculation about poetry and rhetoric. Kenner, that is, restores the promise of conscious unity and univocity Pound failed to transcribe from *The Poetics:* "'But the greatest thing by far is to be a master of metaphor. It is the one thing that cannot be learnt from others; and it is also a sign of genius, since a good metaphor implies an intuitive perception of the similarity of dissimilars' (*Poetics* XXII). Pound has several times cited this statement approvingly" (PP, 87). Here, in his early book on Pound, Kenner emphasizes the way metaphor, as a kind of intuitional addition to nature, elevates perception to the status of Idea, and thus to equality with such categorical distinctions as similarity and difference. This reading is clearly more classical and formalist than that of *The Pound Era,* and thus more deferential to a neoclassical sense of definition than to figural "process." This revision would seem to violate Pound's practice, both in *ABC of Reading,* which canonizes a discontinuous or differential tradition of marginal and even renegade texts, and in *The Chinese Written Character,* which prizes the ungrammatical aspect of figuration over the ideal of translatability formalized in the grammar and lexicons of Western languages.

It can be charged that Kenner refuses to give Pound's criticism the kind of analytical reading he bestows on the poetry, thereby accepting the commonplace that critical or discursive prose is generically distinct from poetry, at once an inferior language and yet a mode of truth statement. For example, he cites as Pound's own the Aristotelian definition of the four types of metaphor, and then applies it almost uncritically to ideogrammic writing, practically repeating Fenollosa's myth that if Eastern and Western languages are different, Eastern and Western philosophies are not (Zen Buddhism, for example, anticipates Hegel, according to Fenollosa) and that graphic writing is the most *objective* (both concrete and non-arbitrary) way of reperceiving true ideas.[12] Kenner's interpolation of Aristotle is telling. Not only does it systematize Pound's notions, clearly legitimizing these with the authority of the first philosopher of poetics; it also supplements Aristotle's elliptical definition with the longer definition of metaphor and metonymy from *The Art of Rhetoric.* This passage concisely summarizes—while correcting—Aristotle's troubled distinctions; nevertheless, the choice of examples might give one pause:

> Metaphor, as Aristotle tells us in another place, affirms that four things (*not* two) are so related that A is to B as C is to D. When we say "The ship ploughs the waves," we aren't calling a ship a plough. We are intuitively perceiving the similarity in two dissimilar actions: "The ship does to the waves what a plough does to the ground." (PP, 87)[13]

Here the examples seem to drift from the distributive or substitutive function of metaphor to its movement, and perhaps not wholly accidentally, by way of metaphors of writing to those of cultivation and transportation. Ships and ploughs are, with poet's barks, the oldest of commonplaces by which philosophy has represented language. But Kenner's stress on the algebraic clarity of analogy, rather than movement and "swift relations," defines Pound's alleged equation of Aristotle with Eastern graphism as the verification of a universal logic, or the ethnocentric *universal* that Derrida calls "White Mythology."

We can now turn to an alternative notion of the effect of metaphor in Pound's curious mixture of (Chinese) graphism and Western grammatology by reading it in the context of the "new rhetoric" that has agitated literary criticism for the past decade. Take for example Derrida's reading of the place of metaphor in metaphysics, of Homer's ships and philosophy's slippages into rhetoric and poetics. Derrida's criticism of criticism unveils the fundamental disturbances which ensue when metaphor and translation are used as models for literary history. Derrida describes— and inscribes—slippages of metaphor within the text of philosophy. Acknowledging what has always been an embarrassment to philosophers (and perhaps the cause of the poets' banishment from their republics), he notes that "metaphor," which names all (figurative) language, exacerbates semantic slippages because it can only be addressed in its own terms— *more metaphorico*. For example, in "The *Retrait* of Metaphor," an essay that explores the "economy" and "usury" of metaphors in not un-Poundian ways, Derrida allows his own argument to be "taken off course," or "transported," from the classical philosophical approach which would categorize metaphors into a figural story of transportational metaphors, or metaphors of metaphor. He proceeds not by analysis or categorization, but by the very slippages—or puns, metonymic substitutions, and so on— that will become his own nondescription or nondefinition. His argument is that there is no meta-metaphor. He strategically translates (*translatio*) "his own" metaphors from several languages, borrowing from Aristotle, Heidegger, Paul Ricoeur (or from a kind of phenomenological anthology) in order to show that metaphor has troubled metaphysics from its beginnings.

In this way Derrida poses the questions (even more radical than Pound's "What is language?" "What is literature?") that motivate the deconstructive enterprise: "*Qu'est-ce qui se passe, aujourd'hui, avec la métaphore? Et de la métaphore qu'est-ce qui ce passe?*" Then he begins—as if automatically (and auto-metaphorically)—to write of autobuses, recalling obliquely Aristotle, Homer's ships, and even Pound's electrical circuits:

> [Metaphor] is a very old subject. It occupies the West, inhabits it or lets itself be inhabited: representing itself there as an enormous library in which we would move about without perceiving its limits, proceeding from station to station, going on foot, step by step, or in a bus (we commute already

with the "bus" that I have just named, in translation and, according to the principle of translation, between *Übertragung* and *Übersetzung, metaphorikos* still designating today, in what one calls "modern" Greek, that which concerns means of transportation). *Metaphora* circulates in the city, it conveys us like its inhabitants. . . . We are in a certain way—metaphorically of course, and as concerns the mode of habitation—the content and the tenor of this vehicle: passengers, comprehended and displaced by metaphor.[14]

Even if we stop short of indulging in the associations of stations with metros, as in "In a Station of the Metro," we are compelled to notice that Pound similarly yokes metaphor, transit, and translation in, for example, *ABC of Reading*'s revision of his earlier Imagist formulations. A definition advanced willy-nilly by means of recurrent metaphors emends Pound's own canonical definition of the Image. The following passage contains several architectonic or textual metaphors, at the same time insisting on the dangerous merging of *stasis* (or theory) and *praxis* that disturbs his "reading":

> The defect of earlier imagist propaganda was not in misstatement but in incomplete statement. The diluters took the handiest and easiest meaning, and thought only of the STATIONARY image. If you can't think of imagism or phanopoeia as including the moving image, you will have to make a really needless division of fixed image and praxis or action. (ABCR, 52)

Like Derrida, who uses several names for his principal deconstructive words which are also concepts and metaphors ("différance," "supplement," "dissemination" all name the unthinkable metaphoricity and infinite translatability of language) that distend concepts, Pound assigns a name, "phanopoeia," to what already has several names—"Imagism," "Vorticism," "interpretive metaphor."

Phanopoeia is the middle term of Pound's three-part coinage—*melopoeia, phanopoeia, logopoeia*—that renames the classical trivium of rhetoric, grammar, and logic. Hence phanopoeia can be seen as a metaphor for a certain kind of linguistic effect which displaces the old categories. It cannot be divorced from rhetoric, or from graphism, or from his "moving" Image. Not only does Pound write metaphorically of metaphor, but his figures for figures are unarguably catachreses, far-fetched metaphors, or, as Derrida defines this indefinable trope: "*a sign already assigned to a first idea should be assigned also to a new idea which has no longer a sign at all, or no longer has a sign as its proper expression*" (WM, 57). Conversely, catachresis abusively applies several signifiers to an "old idea." This renaming of "Image" is characteristic of Pound's definitions, which are compendia of self-(mis-)quotations. But, as Nietzsche and Derrida have taught us, such disturbances can be traced back at least as far as Aristotle's difficulties in charting the movements of language (which is to say "metaphor").[15]

In a rereading of Aristotle necessitated by modern philosophy's (or, more specifically, Martin Heidegger's) simultaneous dependence upon and dismissal of metaphor, Derrida notes the virtual impossibility of separating metaphor from the metaphysics it might illuminate: "The defined is therefore implicated in the defining agent of the definition. As is self evident, there can be no appeal here to some sort of homogeneous continuum, one which would ceaselessly relate tradition to itself, that of metaphysics as well as that of rhetoric" (RM, 15). Metaphor does not lend itself to a history and cannot be hypostatized into a linear or circular representation of tradition. Tradition, then, is only a figure (or an interpretation) and thus a sign that the Truth it contains or conserves is already displaced in language, or in certain kinds of interpretation. With regard to tradition, and even a critique of traditional readings, Derrida connects reading with metaphor, perhaps more comprehensively but nevertheless in a manner not wholly unlike Pound:

> I would be the last to reject a criticism under the pretext that it is metaphoric or metonymic or both at the same time. Every reading is so, in one way or another, and the partition does not pass between a figurative reading and an appropriate or literal, correct or true reading, but between capacities of tropes. (RM, 16)

Almost without exception, Pound cites texts rich in tropes which subvert—or produce a detour on the way to—a total or final reading, or a totalitarian interpretation. He offers a notion of "reading" that, despite his condemnations of rhetoric, is comprised of rhetorical strategies and figures. This effect of reading, apparent in Pound's cross-discursive borrowings of models for literary criticism, was already in force in Aristotle's reading of *The Odyssey*. Derrida gives the following account of Aristotle's attempt in *The Poetics* to lay to rest the already "old topic" of metaphor:

> Metaphor—Western also, in this respect—is retiring, it is in the evening of its life. "Evening of life," for "old age": this is one of the examples Aristotle chooses in his *Poetics* for the fourth type of metaphor, the one that proceeds *kata to analogon;* the first, which goes from genus to species, *apo genous epi eidos,* having as an example, as if by chance " 'Here stands my ship' (*neus de moi ed esteke*), for to be anchored is one among many ways of being stopped" (1457b). The example is already a citation from the *Odyssey*. In the evening of its life, metaphor is still a very liberal (*genereux*), inexhaustible subject which cannot be stopped. (RM, 9)

Derrida's point is that as far as one reads back through philosophy to find a place where language can be grounded in nature (*physis*) as thing or Idea and its image (*eidos*), or even in a metaphorics bounded by definition (*hypotyposis*), one finds metaphor treated metaphorically. This reading escapes the nostalgia of Pound's "ideograph," which simply prefers the poetic to the philosophical, but Derrida's target is the later, nostalgic Heidegger who posits the "proximity" of (poetic) language to Being in the poetic Time of being. Derrida's critique demonstrates that Heidegger's methodological difficulties, his discovery that metaphor is no less inexhaustible—and no more subject to his own ontological claims—than it was at its presumed origins, belie his subtle attempt at once to complete and to overcome metaphysics. While he does not presume to undo metaphysics or to abolish literary history, Pound sometimes happens upon the inextricable matrix of metaphor and interpretation, of poetry and criticism (or philosophy). He generates out of these compound crossings a reading of the tradition that cannot be addressed by his canonical critics, let alone contained by the poetic canon Pound himself felt so ambivalent about.

III

Pound introduces both "How to Read" and *ABC of Reading* by announcing his desire to found—not to question or deconstruct—new methods and curricula for the study of literature. Literature, therefore, never ceases being his privileged category, if only by exceeding all categories. As has been suggested, though he is perhaps unaware of the history of philosophical failures in this regard, Pound chooses what he claims is a method which will augment the metaphoricity of literature and all language. Against the compartmentalizing of "its [Germanic philology's] study was so designed as to draw the mind of the student away from literature into inanity" (LE, 20), he offers revolutionary "disorder," a productive chaos in which metaphors of force and electrical power come into play: "Literature as . . . *nutrition* of *impulse*. This idea may worry lovers of order. Just as good literature does often worry them. They regard it as dangerous, chaotic, subversive. They try every idiotic and degrading wheeze to tame it down" (LE, 20–21).

Throughout his reading text, Pound challenges the categories and hierarchies of canonical literary criticism and linguistics, but in so doing he ironically subverts any hope ultimately to arrive at a comprehensive method for reading, or to establish a proper literary canon that might yield the perfect State as the product of the highest Kulchur. At a point usually deferred for future generations of readers, Pound hopes that, in de Man's formula, he will one day be able to offer interpretative metaphors and a concomitant methodology more exact and natural than scientific investigation and comparison. This is to say that at the end of the long, digressive list of proposed readings, a curriculum more distinguished by its lacunae and omissions than by its inclusiveness, he offers, rather than System, a heuristic canon comprised of dissociated fragments and composed, in one of his favorite metaphors, as a bookkeeper's "loose-leaf system" (LE, 18; ABCR, 38 et *passim*); or, more accurately, as a ledger in which old accounts can be reassessed or reimported ("made new") into the reading economy. Nevertheless, his own tropes, to say nothing of his equation of "natural language" with metaphor and figure, including pictograms, resist any final imposition of order or authority, whether from past literature or modern polemics. This double bind, which traps him between a traditional or conservative notion of literature and his enthusiasm for its anarchical force, can best be seen in his metaphors of electrical transmission and linguistic transformation. For example, when he seems to suggest that literature is itself metaphorical, and not simply composed of metaphors, that literature is a field which effects a "swift perception of relations," he can be said to argue for either a continuity between texts (metaphor as similitude and proportion) or a discontinuity in which one text tropes or transforms a previous one. His attempt to "make it new" would seem to advocate the latter, against the prevailing critical opinion that still links his notion of reading to Eliot's "Tradition." More precisely, it is not whether he embraces one or the other of two incompatible or contradictory positions, but that he shows how one is always inscribed in the other. Moreover, the one is irreducible to the other, setting off the very play of inventive disruption or linguistic energies that allows Pound to define literature as both order and chaos.

Pound's commitment both to the freedom of literature and to the authority of method, even of taxonomy, is most apparent where it has been least remarked; that is, in his hierarchies of poets. In valuing poets according to their individual contributions to a universal canon, according to the parts they play in the "cultural plenum," Pound seems to employ the logic of fixed categories and abstractions he everywhere derides. There is no reason to suspect irony behind his ranking of poets by their "inventions" and by the "verbal" or interpretive force of their metaphors. But these criteria, with their attendant stress on transformation and translation, undo all classification from within. If Pound's irony is equivocal, as much directed against his own authority as the targeted authoritarianism of literary history and convention, it nonetheless works against systematic expectations. More than the revolutionary polemics that surface now and again in his reading text, his idiosyncratic and metamorphic categories disturb canonical literary history, which can accommodate disagreements over "facts" and "interpretations" but cannot tolerate interrogations into the stability of such oppositions. Yet Pound's reading lists and histories continually focus on such fundamental distinctions as that between reader and writer, between poetry and interpretation.

Thus Pound's six classes of persons who have charged literature with meaning (see LE, 23) promise the stability, predictability, and iterability required of codes of style and all other taxonomies. Or so they have been read and applied by Pound himself, when, for example, he claims that "if a man knows the facts about the first two categories, he can evaluate almost any unfamiliar book at first sight" (LE, 24). But his literary historical revisionism shows no mercy to his own categories; in fact, "inventors" and "masters," the first two, are especially problematic. Notwithstanding the graphic similarity and some repeated key phrases of the various appearances of the charts, "How to Read" and *ABC of Reading* present several different versions of inventors and masters. Not only do these categories undergo a rereading between texts, they are unstable from their first articulation. For example, here in a section of "How to Read" candidly titled "Part II: Or What May Be an Introduction to Method":

> (a) *The inventors,* discoverers of a particular process of more than one mode and process. Sometimes these people are known, or discoverable; for example, *we know, with reasonable certitude,* that Arnaut Daniel introduced *certain methods of rhyming,* and we know that *certain finenesses of perception appeared first in such a troubadour or in G. Cavalcanti.* We do not know, and are not likely to know, anything about the precursors of Homer. (LE, 23; emphasis added)

Inventors are clearly necessary to Pound's—or any—literary history, which is at great pains to discover and privilege the first instances of particular forms or styles, no matter how "minor" their writer. Yet not only is he uncertain, despite three appearances of "certain," about the proper names ("such a troubadour or in G. Cavalcanti"), the exact dates, and thus the relative originality of inventions, he also suggests that with further reading Homer, the traditional origin(ator) of Western poetry, might assume a subsidiary or "belated" role. Homer becomes simply a generic name for the reader (also a writer!) whose pre-texts are unavailable to literary history; thus, at its origin, Western culture is secondary and interpreted. Furthermore, "masters," Pound's second category, includes those "inventors" who also synthesize others' inventions. Which is to say, Pound's methodological aspirations aside, these merged categories are not "real" or "proper" measurements. Indeed, the exalted category of "masters" is not only doubled within itself, and therefore hardly "homogeneous," since mastery as he defines it is belated to itself:

> (b) *The masters.* This is a very small class, and there are very few real
> ones. The term is properly applied to *inventors who, apart from their own
> inventions, are able to assimilate and co-ordinate a large number of pre-
> ceding inventions.* I mean to say they either start with a core of their own
> and accumulate adjuncts, or they digest a vast mass of subject matter, *apply
> a number of known modes of expression,* and succeed in pervading the
> whole with some special quality or some special character of their own,
> *and bring the whole to a state of homogeneous fullness.* (LE, 23; emphasis
> added)

Leaving aside the contradiction of "special quality" that comprises a "homogeneous fullness," which is itself a compendium of "some special qualities" or inventions, masters are remarkable for their interpretive capacity. The more material they can read—in order to write—in the greatest number of "processes," the greater their "master-pieces" will be.

It is clearly not the case that masters convey their inherited tradition intact; instead, they continually reread, translate, and transform it. *The Cantos* is exemplary in this regard, but the theme of interpreting fragments can also be seen in Pound's first "long poem," "Near Perigord," which reads Bertran de Born's conventional *Dompna Soiseubuda,* a "composite," "textual," literally "borrowed" lady *composed* out of fragments of love songs, as an emblem or privileged figure of Late Medieval poetry, a metaphor of interpretation. In this way, Bertran de Born, a minor precursor to Dante, invents a metaphor for Pound's literary history, which often privileges such unaccountable fragments. Moreover, Pound's definition of masters, which questions the homogeneity it ostensibly promises, allows that modern writers, if only by virtue of their belatedness, have an advantage over the ancients. Modernism is thus neither a completion of tradition nor a return to origins but is at once new and original because it is both inventive and interpretive—form as well as "trans-form." This is why Pound's masters are not individual or isolated selves but centers of action and points of translation.

If "How to Read"'s revaluation of tradition and innovation, of writer and reader, is equivocal, *ABC of Reading* further complicates the "same" categories when it presents them in condensed form and without benefit of explanation or example:

> 1. Inventors. Men who found a new process or whose extant work gives
> us the first known example of a process.

2. The masters. Men who combined a number of such processes, and who used them as well as or better than the inventors. (ABCR, 39)

Here Pound suggests that by drawing upon previous *texts,* on "processes" rather than poems or other "monuments," masters can "improve" the tradition. Such *improvements* are, one might note, impossible in Eliot's organic and genetic tradition, which is complete and perfect at any given moment, and hardly allows the possibility that modern interpretations and translations could rival, let alone "master," the original. Moreover, if masters "combine" as well as refine, they cannot be at the genesis or origin of anything, but only at a center where a change or alteration of force takes place—a turning, or troping.

Pound's own cross-discursive borrowings and his preference for the polymaths of literary history make clear that masters can affect Poetry more profoundly than those who simply invent styles. For example, Dante, Pound's exemplary master, uses the stylistic inventions of Arnaut Daniel and Guido Cavalcanti in his own epic which also draws upon the political, religious, and economic "news" of the day. Pound's recuperation of Daniel and Cavalcanti, among other "minor" poets who had created something "new," is well documented. Yet even here it is unclear whether inventors take precedence over masters, whether Pound wants to complete or overturn the canon. Under the name "Dante," then, one category slides into another, and one moment of history is textually marked as both an exemplum and a radical break. Seen from one perspective in literary history, Dante is clearly a master; yet he is but one of the inventors upon whom Pound calls for *The Cantos*.

Moreover, if the precise designations and relative values of Pound's favorite Renaissance figures are uncertain, this ambiguity is multiplied in the case of the two principal poets of what he called an American Risorgimento, Eliot and Pound himself. Indeed, the two modern poets might be thought to repeat or trope the exchanges of poetic forces of that earlier chapter in literary history. Pound's apparently objective categories could be said to answer *The Waste Land*'s dedication to Pound, "*il miglior fabbro*" (the greater craftsman), Dante's epithet for Arnaut Daniel, whom he cited as inventor of *The Divine Comedy*'s *canzoni* form. Eliot's tribute would place himself in Dante's role, a position consistent with his own hierarchical "tradition," which prizes continuity over innovation. Thus, not without differences, Pound and Eliot would tend to agree that masters are more valuable than inventors, though each poet claims for himself the higher position. Pound's own frequent reflections back to the glory days of Imagism complicate the ownership of *The Waste Land* and again call into question the utility of his categories. Pound takes credit for inventing the form—and stating the "Do's and Dont's"—Eliot would adapt for *The Waste Land,* a poem Pound acknowledged the "modern master-piece" and a pre-text for *The Cantos,* even though Pound's poem was inaugurated before *The Waste Land* was conceived.

Following Pound's own reading of the Imagist revolution, Kenner finds no problem in naming Pound an inventor and promoting Eliot and Joyce as the canonical modern masters. He simply adopts Pound's distinctions as though they were unambiguous and permanent:

> The inventors: In our time, pre-eminently Pound
> The masters: Eliot, Joyce
>
> (PP, 28)

It should be kept in mind that Kenner's hierarchy appears in *The Poetry of Ezra Pound,* published in 1951, a time when Eliot was the standard of, and Pound marginal to, canonical modernism. Kenner inserts Pound into the curriculum by introducing him to "the reader . . . who knows, for instance, his Eliot and his Joyce—but whom thickets of misunder-standing have hitherto kept at a distance from the poetry of Ezra Pound" (PP, 13). By the time of Kenner's second book-length study, and due largely to his own journalistic and pedagogic championing of the writer of *The Cantos,* Pound stands as "master" of his era, writer of the "pal-impsestic" or "nodal"—all Pound's words—masterpiece that synthesizes sculptural, musical, poetic, even mathematical and scientific "inven-tions." Thus Kenner, whose later interpretation has become normative for "the Poundians," silently repeats the transfers of power that at once define and disrupt literary history. As inventor of Imagism and master of the masters quoted in *The Cantos,* that poem which will confirm the mas-tery of Eliot and Joyce, Pound, according to Kenner's history, engenders himself and then gives birth to the canonical "Pound era."

Despite Kenner's revisions, the case of Pound and Eliot is far from clear, as is the extent to which their rivalry motivated Pound's revisions of literary historical categories. Yet Pound is either unwilling or unable simply to repeat his hierarchy, and thereby to found a "method" of can-onization—one which might place himself at the end of a countertradi-tion, as Eliot culminates a genealogy he had himself established. For ex-ample, a late fragment, somewhat ironically titled "And the True History," offers revisions that might incorporate Kenner's interpretation(s), or even Eliot's renunciation of fragments in favor of an orthodoxy that treats cul-ture as a whole. With a playfulness directed more at literary history than at Eliot, Pound seems clearly to privilege "Masters" over other "prede-cessors," whom he nevertheless finds more interesting:

Literature has been made by four or five or six sorts of writers, and the literary historian may distinguish them as:

A. The "predecessors," inventors, discoverers of fragments which are later used in the masterwork or fused into it. Arnaut Daniel and the better troubadours, the hypothetical ballad writers who went before Homer, etc.

B. The "Great," the "Masters."[16]

Pound, whose reading procedure might nevertheless be faithful to his poetic enterprise, will not commit himself to the minimum requirements of a historiographic or classificatory system, the stable number and definitions of categories that will allow for the construction of a canon. It is only with a good deal of further interpretation, including the choice and rationale of examples and the rather arbitrary imposition of Pound's authority for a final version, that such charts become at all useful. It seems clear, if reservations can illuminate, that the writer of *The Cantos*, a poem which insists on fragmentariness and incompleteness, never univocally devalues "discoverers of fragments" in favor of masters, or revolutionary in favor of canonical writers. In this way, a sort of silent commentary intervenes as a reminder that since there are no pure "nouns" or "verbs" in nature there can be no final interpretations, proper names, or fixed categories in literary history.

Differences between "How to Read" and *ABC of Reading* might indicate Pound's switch from the notion of poetry as force to a notion of poetry as consisting of formal elements—or vice versa—if it weren't for the recurrence in both texts of shape-shifting categories and scientific metaphors for poetry that seem to remark such transformations. Also arguing against progressive conservatism or systematism is the fact that both texts depend on the three-tier hierarchy of melopoeia, phanopoeia, and logopoeia, which analyzes and augments literature's "charge." This supplementary hierarchy, which in the two major reading textbooks follows the six levels of writers, adds complexity to the already untidy categories by focusing on the media—or mediations—between two unstable termini, "writer" and "reader": "Language is charged or energized in various manners" (LE, 25). From a certain formalist perspective, Pound has done little more than rename the commonplace poetic devices, sound (melopoeia), visual imagery (phanopoeia), and association of meanings or ideas (logopoeia). Taken in a wider sense and more seriously, however, this renaming undoes categorization itself: not only does it rival the old (Aristotelian) categories of Poetics, Rhetoric, and Logic, it rereads them by merging the last two categories in logopoeia, which privileges metaphoricity as poetry's unique "untranslatable" thought. But all three categories are internally troubled. For example, melopoeia mixes musical and scientific figures (like electrical force, even telecommunications) in order to update the affiliation of music and poetry: "words are *charged,* over and above their plain meaning, with some musical property, *which directs the bearing or trend of that meaning*" (LE, 25; emphasis added). Music, by this definition, is neither organic nor ornamental; it is instrumental. Like a radio transmitter, it directs energy to communicate an enhanced or, if the signal is weak, a distorted meaning. In the worst cases, one hears only "static" or "white noise" without meaning or purpose. Despite Pound's hope that music, which he says elsewhere should never be far from dance, will direct meaning or "tendencies," his explanation introduces all the potential distortions of radios, and hence entropy.

The errancy of radio transmissions, which is only the enhancement or acceleration of language's inherent "play," is clearest in logopoeia. In a passage reminiscent of Nietzsche's reading of Dionysus's frenetic dance that disturbs ratiocination, Pound defines logopoeia as poetry's transformative force:

> Logopoeia, "*the dance of the intellect among words*" [emphasis added], that is to say, it employs words not only for their direct meaning, but it takes count in a special way of habits of usage, of the context we *expect* [emphasis in text] to find with the word, its usual concomitants, of its known acceptances, and *of ironical play* [emphasis added]. (LE, 25)

Pound's "ironical play" is not to be confused with the New Critical sense of irony as a regulated polysemy fully contained by the poem, which also mysteriously foreshadows the resolution of irony and a resolution of the poem into formal Truth.[17] Logopoeia works against fixed meaning and totalization, things made impossible by the very "nature of language," which, for Pound, is both plural and disseminative. In this sense it is indistinguishable from a certain sense of rhetoric. As we have seen, this notion of irony, not as a simple trope but as another name for linguistic force, appears in Pound's canon at least as early as his reading of *The Chinese Written Character,* where Nature itself is language untrammeled by grammar. Logopoeia, too, is untamable and untranslatable because it is "*trans*lation," the unaccountable and potentially disfiguring movement between languages—or, more radically, what Pound terms the "dance of the intellect" which, while it cannot be stabilized, seems to inhere in the context or scene of reading, as Pound describes it in his most ecstatic moments.

While Pound says that figuration characterizes all language, he reserves the phrase "dance of the intellect" for writing and poetry; that is to say, logopoeia and phanopoeia (the latter is the play of visual images, Pound's favorite category) are ultimately inseparable; ideas cannot be conceived outside of their inscription, or figuration. This terminological slippage undoes the distinctions between language and literature it was ostensibly undertaken to establish. Thus, in "How to Read," logopoeia is the defining characteristic of literature, because it "holds the aesthetic content which is peculiarly the domain of verbal manifestation and cannot possibly be contained in plastic or in music. . . . *Logopoeia* does not translate; though the attitude of mind it expresses may pass through a paraphrase" (LE, 25). To pass through a paraphrase is not to pass through a transparent medium, any more than a radio transmission can be guaranteed to arrive at a *receiver* (whether listener or radio) free of distortion. But, notwithstanding his focus on the linguistic and electrical media, Pound retains hope for "the *transmittability* of a conviction" (ABCR, 27), for an *ethos* free of rhetoricity.

At several other points in *ABC of Reading,* logopoeia and the other categories of poetic expression can be seen to pass over into general features of language. Thus the familiar, if cryptic, advertisement: "Language is a means of communication. To charge language with meaning to the utmost possible degree, we have, as stated, the three chief means" (ABCR, 63). The subsequent formulation defines logopoeia as the "conducting medium" of phanopoeia and melopoeia; it works by "inducing both of the effects by *stimulating the associations* (intellectual or emotional) that have remained *in the receiver's consciousness in relation to the actual* words or word groups employed" (ABCR, 63; emphasis added). Whether as "ironical play" or "stimulus" to the reader's unknowable remnants of thought, logopoeia augments linguistic power and disseminates associations which are always fleeting.

The play on readers, writers, and radios (also ratios) surfaces at other times in Pound's reading text, only to yield an antisystematic "system" that is either self-regulated or, what might be the same thing, unregulated. Pound recognizes the need for standards and measurements, for "conductors" and "voltmeters" (sometimes spelled voltometer) and "antennae"—all Pound's names for the poets who receive and transmit "charged language." That is, he assigns apparently normative roles to the poets—and poems—under analysis, as in, for example, his familiar call to linguistic efficiency: "Good writers are those who keep the language *efficient*" (ABCR, 32). The task of charging language, of augmenting the volatility of (readings in) the tradition, distinguishes Pound's poet-readers from Eliot's "privileged readers," those poets, including Coleridge, Arnold, and Eliot, whose critical canons adequately account for their poetry and thereby provide its normative readings. In place of this orderly incorporation of poetry and criticism, Pound's privileged writers would seem to broadcast news that cannot be interpreted identically by any two readers yet is nonetheless mysteriously affiliated with literature:

> What is the USE of LANGUAGE? WHY STUDY LITERATURE? LAN-GUAGE was obviously created, and is, obviously, USED for communication. "Literature is news that STAYS news." . . . I cannot for example, wear out my interest in the Ta Hio of Confucius, or in the Homeric poems. . . . The above are natural phenomena, they serve as measuring rods or instruments. For no two people are these "measures" identical. (ABCR, 29–30)

The contradiction of "natural phenomena" serving as "measuring rods" and "instruments" indicates the tension Pound's terms must but cannot quite resolve as he tries to account both for the organic and the electrical, for poetry's "natural language" and its function as "interpretative literature," which might rival science in assigning and disseminating meaning.

Yet a system in which measurements cannot be repeated or verified has a high potential for randomness or chaos, especially when it measures linguistic energy, itself capable of generating all manner of disruption. Consistency and verifiability are most problematic in Pound's rationale for his heterogeneous reading "Exhibits," fragments more notable for their technical contributions than either revolutionary content or formal perfection. He no doubt delighted in the shock value of "measuring" such dusty volumes as Gavin Douglas's translation of *The Aenead* and Arthur Golding's of *The Metamorphoses* with the instruments of the New Science. More importantly, this contradiction or juxtaposition of ancient texts and innovation commits Pound to a notion of literature defined as "language charged to the highest degree" which is nevertheless a regulated field (the "rose in the steel dust") of energy. Poetry, that is, cannot be thought apart from a "science of poetry":

> Certain verbal manifestations *can* [emphasis in text] be employed as *measures, T-squares, voltmeters,* or can be used "for comparison," and familiarity with them can indubitably enable a man *to estimate writing in general, and the relative forces, energies and perfections or imperfections* of books. [emphasis added] . . .
>
> The authors and books I recommend in this introduction to the study of letters are to be considered AS measuring-rods and voltmeters. (ABCR, 87)

Pound chooses texts that carry high voltage and serve as voltmeters to measure one electrical charge by another. Measuring force with force would seem Sisyphean, but, in fact, that is how electricians and hobbyists determine voltage. Just as every measurement indicates a relation of an object against a fixed value (I am six feet tall when the fixed point is my livingroom floor, arbitrarily distinguished from other possible "grounds," like sea level or the center of the earth), voltmeters measure relative voltage. Take, for example, a simple pocket voltmeter used to measure house current. The reading it yields, called standard 120, is a differential figure, indicating the relative voltages of "active" and "neutral" wires. The neutral wire is arbitrarily assigned the fixed value of zero. But measured against another wire, the next-door neighbor's neutral wire, for example, the former zero yields a positive or negative reading. Nevertheless, this second—hence every other—meter reading requires a new zero or reference point, fixed at least for the duration of the reading.

Though he extends and revises the metaphors of scientific measurement by calling his favorite classics "axes of reference," Pound is unwilling to fix a reference point in order to assign stable values to works in the literary canon. Or, what is more abusive of method, he chooses different reference texts as his reading lists undergo expansion and revision. The classics and their writers—or, better, producers, for not all of Pound's exemplary texts are written—are sometimes incompatible. Pound admits that, because their works are "verbal" and "interpretive," artists cannot be fixed, as proper nouns, in catalogues: "Any sort of understanding of civilization needs comprehension of incompatibles. You can't Goya and Ambrosio Praedis at the same time, using those names as verbs" (GK, 184). Without fixed references and proper names, the "swift relations" among and within texts are anamorphic and unmeasurable. This condition of reading is exacerbated by the highest classes of writers, inventors and masters, all producers of "interpretive metaphors" that continually undergo rereading. For instance, the most perfect master of *The Spirit of Romance,* a text addressed to literary specialists, becomes a "foreign stimulant" in *ABC of Reading,* a textbook for "those who might like to learn." Dante's ouster by Chaucer, which involves more than the translation of the continental Renaissance into English, is an obvious, if neglected, case in point.

The "Exhibits" section in *ABC of Reading* affords Chaucer, that ambiguous figure so dangerously marginal to French thought and English poetry, a heretically high position in the English literary canon (especially at a time when Eliot had hypostatized the Renaissance intellectual lyric) as its transitional or transformational moment: "Chaucer wrote while England was still a part of Europe. . . . Chaucer's name *is* French and not English (ABCR, 100–101). Perhaps because readers naturally pass quickly over Pound's ninety-nine-page compendium of the high points of Western literature to less quirky anthologies, no one has found his reading of "Master Geoffrey" particularly noteworthy. Yet it does overturn Pound's familiar readings of the Late Middle Ages and Renaissance. Interestingly, Chaucer also marks the slippage between literary historical epochs, writing as he did during the English Middle Ages, which lingered while the Italian Renaissance was well underway. Moreover, in the course of a few pages, Chaucer surpasses Pound's usual poet-readers, Dante and the Troubadours, to complete the contintental and found the English canons. At its beginnings, then, since Pound omits the Anglo-Saxons, English poetry's greatness rests on interpretation and translation: "Chaucer wrote when reading was no disgrace," and, in a phrase which makes him very much "Pound's Chaucer," he "was a man with whom we could have discussed Fabre and Fraser [sic]" (ABCR, 100). Although Pound does not say so, at least not in so many words, Chaucer falls into the gray area between masters and inventors. On the one hand, he composed the elements of a unified culture out of a compendium of (heterogeneous) fragments and translations: "There was one culture from Ferrara to Paris and it extended to England. Chaucer was the greatest poet of his day. He was more compendious than Dante" (ABCR, 101). On the other hand, Chaucer invented both a national poetry and his own mother tongue (which is the English, not the *Amerikun,* idiom): "He is Le Grand Translateur. He had found a new language, he had it largely to himself, with the grand opportunity. Nothing spoiled, nothing worn out" (ABCR, 101). Of course, by Pound's own definition, Chaucer was at the end of a great tradition; his "new language" is at least trebly translated—from the Greek and Latin classics through Dante's Italian and the French of Guillaume de Lorris and Jean de Meun to his own vulgar tongue, English. But Pound preferred his classics second-hand—like translating his *Odyssey* through Divus—and his translations abusive, as early as "The Seafarer" and "Sestina Altaforte."

The complex formulation of an "originary translation," of a poetic force augmented—not diffused or worn down—by reading and translation, and thus working an exchange of roles between inventor and master, is repeated and revised over the course of Pound's reading career.[18] In opposition to canonical readings, which trace a continuity among texts within the tradition, Pound's readings do not disclose a pure or proper Homer or Dante, available in primordial or original texts. Instead, he charts a number of translations and appropriations that are signified, not by proper names, but by possessive adjectives and generic names. Thus he reads—and writes of—"Divus's Homer" or "Gavin Douglas's Virgil" or "Binyon's Dante" or even "Pound's *Cantos*" (the latter cross-references, but does not *sublate,* the former translations). In affording Chaucer, as "Le Grand Translateur," a position equal to writers of "original" English poetry, Pound suggests that the "force" of poetry, the "charge" and "change" undergone in the course of translation, is more important than formal poetic elements. This is not to say, however, that Pound contends himself with the abyssal play of languages such a focus implies. On the contrary, he insists that certain details and "elements" remain intact or at least identifiable amidst these "swift relations."

In other words, slippages within and among his various lists, charts, and exhibits that fail to cohere into one method are exacerbated by Pound's changing characterization of the objects under analysis. At times poets or poems are "processes" and "forces"; sometimes they are stable "identities," even "elements." In this way, "How to Read" would seem to go off course from the beginning by introducing stable definitions of literature in electrical metaphors: "Great literature is simply language charged with meaning to the utmost possible degree" (LE, 23). Literature compounds the force of languages, even, as we have seen, to the point where its *charge* exceeds the ability of tradition to contain or transmit it. Pound's privilege or "donative" writers increase the volatility of their medium: "this charging has been done *by several clearly definable sorts of people*" (LE, 23; emphasis added). The question again arises of how to determine the origin of poetic force in a fluxional field, for even on a simple electrical wire the source of additional charge is unclear, as the charge is the same along the length of a conducting wire. But Pound insists on having both untamed force and singular or univocal elements. Therefore, *ABC of Reading* advertises one of its unstable hierarchies as a schematic of "pure elements," like the Periodic Table: "When you start searching for 'pure elements' you will find that literature has been created by the following classes of persons" (ABCR, 39). Pound's apparent indecision about whether tradition is a series of privileged ("luminous") details or an indefinable and disseminative "force" is apparent in the early series of essays, "I Gather the Limbs of Osiris," where he suggests that interpretations and translations partake of the (electrical) "power" they would regulate—that certain privileged "facts" are at once the Law and lawless and thus an excess of law:

> Any fact is, in a sense, "significant." Any fact may be "symptomatic," but certain facts give one a sudden insight into circumjacent conditions, into their causes, their effects, into sequence, and law. . . .
>
> These facts are hard to find. They are swift and easy of transmission. They govern knowledge as the switchboard governs an electric circuit. (SP, 22, 23)

In this way, interpretative metaphors direct other metaphors along an endless chain of interpretation. How can a switchboard also be swift, or literature dictate its own interpretation(s)?

Pound's "reading" hardly puts an end to reading or provides us with
a program to govern our readings. The pluralism—the very grammatical
pluralness—of his hierarchies and methods generates more questions than
it answers, more transformations than it can name or tame. This exchange
of (verbal) power and interpretations is charted in his earlier critical works.
For instance, in *The Chinese Written Character* we are confronted by the
rather surprising equation of "natural forces" with reading and writing,
whose agents are fused and transformed by their activity: "The development of the normal transitive sentence rests upon the fact that one action in nature promotes another; thus the agent and object are secretly
verbs. For example, our sentence, 'Reading promotes writing'" (CWC,
29). Characteristically, Pound's example and explanation undo the very
grammar they were advanced to explain. While suggesting the undesir-
ability, even the impossibility, of program and method, such a notion of
language underwrites—or, better, subverts—Pound's reading text, in such
a way that it repeatedly affirms the antisystematic force of poetry his
prescriptive taxonomies seem to name in order to reify.

Almost despite himself, Pound seems to have inculcated Fenollosa's
"method," that had first to disrupt grammar and logic in order eventually
to achieve "valid scientific thought." This natural-scientific-poetic thought
"consists in following as closely as may be *the actual and entangled
lines of forces as they pulse through things*" (CWC, 12; emphasis added).
And Pound frequently invokes the "verbal" and "metaphoric"—the "ra-
dio-active"—power of Chinese writing, equating these with electricity
and other products of modern science. A particularly suggestive passage
in the Pound/Fenollosa essay interjects untamed electrical current into a
notion of organic language and metaphor (rhetoric) into grammar:

> The truth is that acts are successive, even continuous; one cause passes
> into another. And though we may string ever so many clauses into a single
> compound sentence, *motion leaks everywhere, like electricity from an ex-
> posed wire*. All processes in nature are interrelated; and thus there could
> be no complete sentence (according to this definition) save one which it
> would take all time to pronounce. (CWC, 11; emphasis added)[19]

While Fenollosa retains a certain nostalgia for the perfected sentence,
he suggests that Western grammar, which prescribes the workings of sentences, is *contra natura,* and so are the methodologies it grounds. Thus,
a passage later in the same essay reads the nineteenth-century revolt against

the taxonomic system as a return to a more expressive—or forceful and metaphoric—language:

> It is impossible to represent change in this system or any kind of growth.
> This is probably why the conception of evolution came so late in Europe.
> *It could not make way until it was prepared to destroy the inveterate logic of classification.* (CWC, 27)

Fenollosa's reading of Darwin, and his call for a return to the Chinese graphic language, would seem simply to reverse the "priorities" in which stasis is posited above and before flux, idea over interpretation, thought over poetry; but binarism and simple hierarchization remain from the old— or "natural"—sciences and Germanic philology. The case of Pound's later tamperings with the "logic of classification" are more complex, and Fenollosa is but one precursor to his ambiguous reading enterprise.

In his ambition to found a method at once prescriptive and faithful to poetry's metamorphic force, Pound also calls upon Louis Agassiz, the famous Harvard zoologist and geologist. The same passage that advocates Fenollosa's "non-method," in fact, contains the anecdote of Agassiz and the fish. Pound retells the story of the graduate student who observes (but does not see) a fish so long that it rots before he can completely describe it in the arcane Latin taxonomical names at his disposal. Pound seems tempted to ask, if we might for a moment indulge the terms of current reader-response criticism, "Is there a fish in this class?" The point is that the student, like readers of poetry, should go directly to the fish or text and describe its system of interrelated parts rather than apply received categories:

> The proper METHOD for studying poetry and good letters is the method of contemporary biologists, that is careful first-hand examination of the matter, and continual COMPARISON of one "slide" or specimen with another. (ABCR, 17)

The analogous *Idea*—or, better, anti-ideology—of reading poetry without the interference of nonliterary interpretations is so basic to New Critical theory, which claims Pound as a forebear, that the contradiction of interfiling Fenollosa and Agassiz has escaped notice.

In juxtaposing, or nearly superpositioning, those figures, Pound has charted his methodological double bind. If Fenollosa represents an infusion of the poetic and metaphoric into science, Agassiz represents the reaction of natural science against such revolutions. Agassiz is indeed remembered for his advocacy of "direct observation," a phrase of empiricism, which he inherited from his teachers, Von Humboldt and Cuvier. And Pound is again correct; he did observe fish, but long-dead ones.

His major work on the subject is *Recherches sur les poissons fossiles* (5 vols., 1833–1844). But by far his most influential work is "Essay on Classification" (see *Contributions to the Natural History of the United States*, 4 vols., 1857–1862), whose rigid taxonomical system argues for a notion of fixed or geologic time, against evolution. In fact, Agassiz is most remembered for his insistence, in direct opposition to Darwin, that taxonomy can adequately, and once and for all time, define biological species and races. Pound could not have been unaware of Agassiz's program and reputation. Moreover, his own persistent, if half ironic, belief in the pedagogic and evaluative utility of rigid classificatory systems could easily have attracted him to Agassiz's taxonomies. Pound's disruptions of fixed categories by "poetry's charge," his sense of nature (and language) as accounting for transformations within and across the strict demarcations of closed categories, in short, his sense that natural poetic language, as Fenollosa said of the Chinese written character, reintroduced Time into classical notions of poetry, destabilize any notion of science. Agassiz and Darwin are curiously entwined in Pound's version of the new science, forming an incompatible opposition or *aporia* of methods which render the missappropriated "proper" names of two sciences or scientists.[20]

Indeed, the search for a "dynamic system," for a strict method that might nevertheless accommodate change and revision, occupied Pound's whole career, surfacing in the various artistic *movements* to which he belonged, including Imagism. Unlike many of his contemporaries, Pound did not renounce his early experimentalism; rather, he erected experiment as a systematic ideal and movement as a necessary motive for art. For example, the passage from *ABC of Reading,* cited above, can now be seen to have wider methodological implications:

> The defect of earlier imagist propaganda was not in misstatement but in incomplete statement. The diluters took the handiest and easiest meaning, and thought only of the STATIONARY image. If you can't think of imagism or phanopoeia as including the moving image, you will have to make a really needless division of fixed image and praxis or action. (ABCR, 52)

Thus, even as he (re-)canonizes his own earlier enthusiasms, Pound emphasizes a poetics that "works" in metaphors intended to read the tradition. It is no accident, then, that he seems particularly drawn to those writers similarly caught in this double bind between conventional systems or vocabularies and experimentalism.

By the time of *Guide to Kulchur,* Pound's desire to chart both forces and details found a compatible model in Leo Frobenius's *Kulturmor-*

phologie, a discourse from which he borrowed "Paideuma" and the "kinematic map." Like Fenollosa, Frobenius wanted to recover and represent
"verbal force." If his field was an anthropology "not retrospective but
immediate," rather than linguistics itself, his focus on scientific language
still involves the *translation* of force into a new, if questionable, "method"
and vocabulary. Thus Pound gives the following account of the coinage
of "Paideuma," a notion that draws upon *Paidia,* a Greek word covering
play, in general, and linguistic figuration in particular: "To escape a word
or set of words loaded up with dead association, Frobenius uses the term
Paideuma for the tangle or complex of the inrooted ideas of any period"
(GK, 37). Characteristically, Pound is attracted by the transformation of
an ancient word into a revolutionary method. Particularly attractive is a
neologism that yokes ancient Greek, the language of the first (Aristotle's)
science and the new science of anthropology. Frobenius's alleged terminological revisionism, his rejection of old for new associations, which
will receive full treatment in Chapter Five, seems to confirm Pound's own
reconsiderations of Imagism.

Frobenius's "kinematic map" of African culture, which Pound transposed into *The Cantos* as the organizing metaphor, *periplum,* a map of
shores as sailors might see them from aboard ship (a mistranslation of
Hanno's *periplus* and an echo of Homer's [and Derrida's] ships), epito-
mizes the war between "facts" and "interpretations"—between "swift relations" and the maps or taxonomies that chart them in order to give each
particular its place. Hence the following translation from Frobenius's
German:

> Culture is a living reality which can only be comprehended by penetration
> of facts. . . . *The image must be nothing more than a means.* The problem
> is to represent movement which can only be done by *deflecting attention
> constantly from the static factual to the genuinely kinematic.* . . . Since
> culture is kinematic, the map must become kinematographic. In recording
> the movement of culture in a series of maps we avoid hypotheses and *ren-
> der the invisible to the searching eye as "map reading."* (emphasis added)[21]

This attempt faithfully to represent movement radicalizes the place of
interpretation, placing the reading process prior to fact and perception
(seeing). Such a map, which works only by indirection, "by deflecting
attention . . . from the static factual," implies a criticism or reading of
maps. While Frobenius aims to avoid the mediations of science and logic
("facts" and "hypotheses"), his switch from "invisible" data to "visible"
representation (which is, incidentally, Aristotle's most familiar and logical definition of metaphor) involves "map reading"—or, look again,

"reading map reading." This hardly confirms the unimpeded transfer of
action implied by the phrase "the genuinely kinematic." Furthermore, the
subject under analysis and the analytic method are both called reading.
According to this model, therefore, there can be no thing (Ding-an-sich)
or Truth prior to reading. Thus are compounded the problems Derrida
finds attendant upon philosophical discussions of metaphor, *more metaphorico:*

> The defined is therefore implicated in the defining agent of the definition.
> As is self evident, there can be no appeal here to a sort of homogeneous
> continuum, one which would ceaselessly relate tradition to itself, that of
> metaphysics as well as that of rhetoric. (RM, 15)

Neither rhetoric nor philosophy can account for the slippages of metaphor
among philosophy and literature and science. Similarly, Pound's notion
of "interpretive literature," which draws upon the most problematic figures of all three discourses, cannot be said to found or ground a modern
tradition. Pound's borrowings from a French/English translator–poet,
Chaucer, and a neo-Nietzschean/Spenglerian anthropologist–art historian, Frobenius, suggests the lengths to which he was willing to go in
order to "make it new"—or, better, to answer the questions, "What is
Literature?" "What is language?" etc. They do not comprise a tradition,
but reassemble, and in Charles Olson's figure of "projective" verse, virtually "project" one.

IV

Regardless of the sources of such lexical and methodological disruptions,
Pound insists that reader and writer (should) contribute additional force—
"positive charge"—to the already volatile media of literature and language. In a passage remarkable for invoking the very revolutions it dismisses, he says: "Thought was churned up by Darwin, by science, by
industrial machines, Nietzsche made a temporary commotion, but these
things are extraneous to our subject which is the *art of getting meaning
into words*" (LE, 32).

Notwithstanding Pound's dismissal, Nietzsche provides the most thorough account of the power of metaphor and interpretation and their place
in the reading economy. According to Nietzsche, the mission of the modern artist (and of the "Dionysian philosopher") is to reform language. His
early aesthetic writings, which, incidentally, found receptive audiences
in Remy de Gourmont's Paris and A. R. Orage's London and thence
filtered through "the air" surrounding Pound's early works, praise Wag-

ner as the "perfect Dionysian," for marrying the incompatibles of music, drama, thought, and emotion in a new artistic language liberated from philosophical constraints. Thus Wagner's importance lies in his attention to language as such and to all artistic media: "He was the first to recognize an evil which is as widespread as civilization itself among men: language is everywhere diseased" (RW, 132). Indeed, before succumbing to Hegel's System, Wagner allegedly used a kind of "interpretive metaphor": "For a genuine poet metaphor is not a rhetorical figure, but a vicarious image that he actually beholds in place of a concept" (BT, 63). An echo of that substitution of poetic for abstract thought is unmistakable in Pound's formulation of "interpretive metaphor" as a tropology free of the ornament and deception of ordinary rhetoric and more effective than philosophy. His reading of *The Divine Comedy* as "single elaborated metaphor of life" (SR, 87) is exemplary.

Nietzsche's reading of the history of philosophy goes further: he comes to question the reification of art and the mutual exclusivity of poetry and philosophy. Yet, even after Wagner had fallen in his estimation, after "art" ceased being the exclusive category to which philosophy should aspire or return, Nietzsche associates language, art, and interpretation with power, in the endlessly generative and interpretive Will to Power. He does not hesitate to destabilize all system—including his own text— by claiming that interpreter and text, reader and writer, are simply "grammatical fictions" in the fluxional field that is the world—without origin, aim, or end, yet endlessly productive:

> This, my *Dionysian* world of the eternally self-creating, the eternally self-destroying, this mystery world of the two fold voluptuous delight, my "beyond good and evil." . . . Do you want a *name* for this world? A *solution* for all its riddles? . . . *This world is the will to power—and nothing besides!* And you yourselves are also this will to power—and nothing besides. (WP, 550)

Here Nietzsche affirms the impossibility of fixing "power" in names or categories acceptable to the history of ideas, speaking instead in riddles and metaphors. Will to Power is not a concept or a representation of a Truth prior to language; instead, it is a metaphor of metaphor. As other aphorisms make clear, it is thoroughly rhetorical. Which is to say, in the terms de Man borrows from the early Nietzsche, it is both tropological and persuasive—in the sense that it is a fiction constructed to advance his analysis.

Nietzsche erects Will to Power as a strategic intervention into those logical and taxonomic systems that attempt to halt the play of interpretation and translation. Far more radically than Fenollosa's Nature, which can still be represented by (a misreading of) the Chinese ideograph, or Frobenius's "kinematic map," a sort of double registration of the "static factual" and "motion," Nietzsche addresses the *"essence* of language"— a phrase made absurd by his demystification of the notion of essence but nevertheless strategic in his critique of the metaphysical effort to go behind language to essence in order to represent Being.[22] Therefore, Will to Power is not a truth statement, because it cannot be thought without challenging that normative commonplace of Western thought named Truth: "It expresses the characteristic that cannot be thought out of the mechanistic order without thinking away this order itself" (WP, 338).

Whereas Pound's mastery over his metaphors of power, and the transformations these mark in the interpretive systems he borrowed, remains ambiguous, Nietzsche would seem to exploit the very "swift perception of the relations of things" Pound took to be an unproblematic definition of metaphor. The relations among, and the radical instability within, the models of the physical sciences and metaphysics are Nietzsche's theme. But not even "perception," before which logic posits a "perceiver"—a fixed perspective and grammatical subject—is left intact when he says that a fixed "Truth" is simply a "nonsensical" linguistic construct. I quote this representative passage at some length, because it suggests the broader theoretical implications of Pound's indecision about whether to analyze poetry's "processes" or "pure elements," the "movements" or "details" of literary history:

> The mechanistic concept of *"motion" is already a translation of the original process into the sign language of sight and touch.*
>
> The concept "atom," the distinction between the "seat of a driving force and the force itself," is a sign language derived from our logical-physical world.
>
> We cannot change our means of expression at will: it is possible to understand to what extent they are mere signs. The demand for an adequate mode of expression is senseless: *it is of the essence of a language, a means of expression, to express mere relationship—*
>
> *The concept "truth" is nonsensical. The entire domain of "true-false" applies only to relations, not to an "in-itself"*—There is no "essence-in-itself" (it is only relations that constitute an essence—), just as there can be no "knowledge-in-itself." (WP, 334; emphasis added)

By isolating such crucial moments—or, better, scenes of reading or translation—at which physics and metaphysics willfully ignore the role of Will to Power as interpretation, Nietzsche questions the very notion of linguistic adequation and the role of Will as constituting Self or originating Power. Indeed, this question remains open for modern physics, a science nearly merged with ontology, which cannot decide whether "matter" is constituted by "waves" or "particles," newer words for the old ideas, motion and atoms. But Nietzsche's interrogation is more radical than that of an older metaphysics which attempted to ground science, because he insists that the binary oppositions of the inquiry are themselves arbitrary impositions onto an "original process." This process remains inaccessible even to perception; for perception, that orienting center of nineteenth-century empiricism, is itself a mere "sign language" and thus subject to the conventionalized arbitrariness of grammar and reading.[23] Finally, "essence" and "origin" are already complex interpretations, translations of "force" into philosophical abstractions that have a long history of interpretation by previous philosophers. Nietzsche does not claim to escape intertextuality or "the arbitrariness of the sign"; instead, he marks reading or translative moments at the very "origin" that philosophers and poets alike have, in their nostalgia, claimed to be prior to language and free of its distortions. Not only does he expose the translations and rhetoricity repressed by philosophy, but he also lays open his own text to translation and (mis-)interpretation. Indeed, his use of two phrases, "original process" and "mere relationship," for the power—or Will to Power, or, in Derrida's Nietzsche-inspired term, "différance"—allegedly prior to the originary translation ("motion") he charts, indicates the inescapability of translation. In this way, Nietzsche's own text evades totalization, refusing to contribute a new or "master" concept—even a "master metaphor"—that might make him, in Heidegger's phrase, "the last metaphysician."[24]

Nietzsche uses Will to Power to interrogate philosophy's fundamental concepts, yet he refuses to grant his coinage ontological status. There is no "Will-to-Power-in-itself" to replace Truth or *das Ding-an-sich;* he reads back through the layers of philosophical interpretation and "error" only to uncover the same problematic of language that motivated his quest. He thus abolishes, by parody, those "genealogical readings" which claim to recover Truth at the "origin," behind an orderly history of ever more reified interpretations. In a passage that has become crucial to current interpreters of Nietzsche, Will to Power names the meaning-giving act in such a way as to expose "Being," that most metaphysical of all abstractions, as a "falsification," belonging to rhetoric, not to physics or to metaphysics:

> To impose upon becoming the character of being—that is the supreme will to power.
>
> Twofold falsification, on the part of the senses and the spirit to preserve a world of that which is, which abides, which is equivalent, etc. (WP, 330)

Here Nietzsche makes impossible the equation of Will to Power and Being by abolishing the very possibility of equivalence. Moreover, he makes explicit the rhetorical nature of Will to Power, which is the imposition of the word and category Being, as well as the halting of temporality. As we have seen, he does not simply displace Being, or any other *first principle,* onto the *subject*—onto the poet who might fictionalize, or the philosopher who might adequately define, Truth. His refusal to halt his reading at Being or becoming, to see reading as a closed economy regulating the exchanges that poetry or music might symbolize, has not prevented readers of Nietzsche from finding such themes—even in this, perhaps his most radical attack on such readings.

For example, as if *willfully* ignoring the second sentence's exposure of philosphical repression—Nietzsche's phrase is "purposeful *forgetting*"—of temporality and interpretation, Heidegger cites the first sentence of that very passage as proof that Nietzsche has stayed within the metaphysical tradition by equating Will to Power and Being. In fact, it is on this passage that Heidegger grounds his canonization of Nietzsche as "the last metaphysician." *Will to Power,* the notes and fragments which became a book only posthumously, becomes for Heidegger the culmination of Nietzsche's canon, which can then be read as a system of equivalent formulations catalogued and stabilized in the book which preserves its author's final words. Heidegger imposes order on Nietzsche's text thus:

> At the summit of the completion of Western philosophy these words are pronounced: "To *stamp* Becoming with the character of Being—that is the *highest will to power*. Thus Nietzsche writes a note entitled, "Recapitulation." According to the character of the manuscript's handwriting we must locate it in the year 1885, about the time when Nietzsche, having completed *Zarathustra,* was planning his systematic metaphysical *magnum opus*. The "Being" Nietzsche thinks here is "the eternal recurrence of the same." It is the way of continuance through which will to power wills itself and guarantees its own presencing as the Being of Becoming. (EGT, 22)

In this way, Heidegger ascribes systematic closure to the very metaphor Nietzsche used to read or undo the dream of "System." Fixing on the section title, "Recapitulation," assigned by Gast, Nietzsche's great disciple and systematizer, Heidegger finds in Nietzsche's most problematic text a "systematic metaphysical *magnum opus,*" replete with clear authorial intention—or "planning"—and a fixed terminology that orders earlier stages on the way to the "Recapitulation," "summit," or "completion" it allegedly achieves. Heidegger manages to change the thrust of Nietzsche's words by a reading that moves among close attention to— almost an "inhabiting" of—Nietzsche's words, and such inventive inattention, or ab-use, and an acceptance of Gast's editing, that results in a kind of interpretation. He ignores the fragmentary state of Nietzsche's text, which, by the way, is not simply the result of his death, his failure to "author" a whole or authorized version, since *Will to Power* was composed and interpolated from more than twenty years of Nietzsche's writing. Nevertheless, Heidegger ascribes "continuance" to the notion of Being, which Nietzsche dismissed as a false "abiding," and thus Heidegger makes the Will to Power a metaphysical principle and a stable term translatable into "eternal recurrence."

Exploring the breathtaking ironies—the double and treble "falsifications"—of Heidegger's reading would necessitate a return through several translations and a journey through the thickets of current French neo-Nietzscheanism, a digression that would take us too far off course. But it should be remembered that Heidegger's is hardly a simple misreading, that he is Nietzsche's subtlest modern reader—which is not to say that he has gone beyond the Will to Power, but that he has grasped its full implications for philosophy and, almost in the same gesture, as Derrida shows, denied them. Let us say now, for the sake of economy, that Heidegger's reading of Nietzsche manages to repress the radical play of metaphor, and hence of literature (poetry) that Nietzsche unveils and that recent Nietzschean criticism has recovered, not simply to resist Heidegger but to open the possibility of a radical criticism or theory of reading.[25]

In this regard, it is noteworthy that Heidegger employs similar procedures to those which literary historians use in establishing a writer's corpus and *his* (not "his or her") place in the canon. By referring to Nietzsche's handwriting, the closest thing to—or *trace* of—the authorial presence, he assigns the fragment a date in order to read it against Nietzsche's published and more philosophically legitimate formulations. Such scholarship (as opposed to more dangerous "criticism") is also a necessary part of reading Pound's critical notes and fragments, which, like Nietzsche's, often appear in several published and unpublished versions of ambiguous authority. Yet fixing the text is a necessary first step to reading, and Heidegger has done a remarkable job with the anamorphic notions and aleatory notes that comprise what can only be "Heidegger's Nietzsche," his own planned *magnum opus*, interpretations of *Will to Power*. One might say that, somewhat in spite of himself, Heidegger engages Will to Power in a most productive way. By inserting his own reading at those points where Nietzsche asserts the rhetoricity of "Will to Power," he functions both as interpreter and interpretand. Heidegger repeats—but in a repetition that marks a difference and indicates the workings of Will to Power—Nietzsche's unthinkable statement that "this world is the will to power—and nothing besides! And you yourselves are also this will to power and nothing besides" (WP, 550; quoted above). In this way, Nietzsche's text "reads" Heidegger's. And, perhaps not wholly fortuitously, Heidegger's text on Nietzsche remains a series of unaccountable lecture notes, essays, and asides within his more canonical works. The condition of his own text and the rigor of his investigation of the workings of Will to Power prevent Heidegger from accomplishing his special mission, which is to go beyond metaphysics and finally to "retreat from language."[26] Instead, the text of *Will to Power* continues to generate more readings out of his own text. Such readings that promote reading are called for by Pound's notion of reading. And Pound's most important readers, sometimes against their best efforts to tame it, have tapped this potential chaos of interpretation—or is it a chaos of interpolative potential, a Joycean "chaosmos"?

If Pound and his canonical readers do not pose generic and linguistic problems in the terms of modern philosophy, similar exchanges of power have taken place over Pound's reading text, especially at those points where, touching upon if not engaging philosophy, it seems to promise a coherent poetics, even a reading program. One might enumerate parallels between Heidegger's troubled reinscription of Nietzsche into the genealogy of metaphysicians beginning with the pre-Socratics and Kenner's reading of Pound's Aristotelianism; some of these have already been considered in connection with the reading of Aristotle that Kenner has ascribed to Pound's notes in *The Chinese Written Character* and *How to Read*. But an even more startling imposition of system occurs with regard to *ABC of Reading* and of what, at face value, seems Pound's most straightforward formula for writing and reading poetry, DICHTEN= CONDENSARE.

In spite of its interpretation by the Poundians as a simple call to Imagistic concision, that "equation" is really an emblem for the subtlest questions of reading and translation, which, following Nietzsche and Heidegger, modern philosophy has located in the differential relation of *Dichten* and *Denken*. Indeed, from a certain perspective, and in the context of Pound's other investigations of reading, translation, and metaphor, his formula refers to "verbal nouns"; it offers a method at once arbitrary and necessary (or natural) and exemplifies the salutary distortions of translations—and all this by way of an accidental "discovery." But before turning to Pound's formula, we should recite Kenner's reading of it.

Kenner translates DICHTEN=CONDENSARE as "verse differs from prose only by superior concentration" (PP, 73). That is, he simply nominalizes both verbs, as a first step to erecting them into a principle or axiom that confirms his own reading of Imagism, as the economic and "objective" use of language. Specifically, Kenner interpolates his reading of "condensare" into Pound's account, in *Gaudier-Brzeska* (1916), of the composition of "In a Station of the Metro." Thus he subtly lends narrative continuity to Pound's poetics as well as to the exemplary Imagist poem. Yet, at the same time, he seems nearly to ignore Pound's renunciation of the static for the dynamic Image, focusing instead on its objective properties (even its "weight" and "number" of words). And, perhaps not accidentally, *Dichten,* with its attendant problems, seems to drop almost entirely from his reading, when he retells Pound's experience of writing his poem and the "experience" behind the poem: "After several decreasingly wordy attempts . . . he succeeded in *boiling away the contingent distraction of the original experience*" (PP, 73; emphasis added). In line with his own "Aristotelianism," his belief that the Image can *adequately* convey the "node" of an experience which is prior to language, Kenner emphasizes Imagism's economy of representation. Then, in a lengthy footnote, ignoring the ways in which Pound's alleged equation questions adequation and representation, he adds:

> This is as good a place as any to refer to 'Dichten=Condensare': Pound's statement . . . that verse differs from prose only by superior concentration. *Its rhythmic and melodic accessories afford extra emotional weight without extra words.* To put it another way, *400 pages of prose may "by sheer architectonics" achieve the weight of 40 lines of verse.* (PP, 73; emphasis added)

If he has perhaps overstated his argument for condensing—after all, he nearly says a pictogram is worth a thousand words—Kenner is certainly not off the mark. Especially in the early criticism, which Kenner tends to prefer, Pound similarly argues for concision. For example, he calls "Hugh Selwyn Mauberley" "an attempt to condense the James novel" (L, 180), and, in a passage from "A Few Don'ts" that concerns the condensed Image, he says, "It is better to present one Image in a lifetime than to produce voluminous works" (LE, 4). Even his reading text is informed by this desire to save words and time, to provide an anthology worthy to be "'A portable substitute for the British Museum'" (LE, 16). Nevertheless, often in the same arguments that emphasize concision or condensing, Pound also mentions the "movement" and "process" that escape or destabilize simple condensation. As we have seen in redirecting attention from the static to the dynamic properties of the Image (ABCR, 52), Pound pays less attention to poetry's concentration than to its force. It must be allowed that Kenner has elsewhere attended to the transfor-

mative and energetic properties of the Image, though he has contained
these in his own metaphors of "Patterned energy" and "nodes."[27] But he
leaves unexplored the interpretive problems of *Dichten,* DICHTEN=
CONDENSARE, and Pound's own troubled dismissal of these.

The longer passage from which critics have extracted the apparent
equation (and some only half of it) turns out to be another instance of
Pound's critical aphorisms undone by the very rhetoricity he finds inherent in all language—and then as quickly forgets:

> "Great literature is simply language charged with meaning to the utmost
> possible degree."
>
> Dichten=condensare.
>
> I begin with poetry because it is the most concentrated form of verbal
> expression. Basil Bunting, fumbling about with a German-Italian diction-
> ary, found that this idea of poetry as concentration is as old almost as the
> German language. "Dichten" is the German verb corresponding to the noun
> "Dichtung" meaning poetry, and the lexicographer has rendered it by the
> Italian verb meaning "to condense." (ABCR, 36)

Since it appears on the page right after the other definition, the formula
should exemplify or repeat Pound's notion of literature as verbal force or
"charged language." That is, especially as the translation must pass through
nominalization to become a statement about poetry, it is a doubly differential formula and not an equation. Pound hardly disguises the fact
that DICHTEN=CONDENSARE is both grammatically and methodologically suspect, trembling with contradictions that fundamentally question
the procedures of literary criticism. Thus Bunting's "fumbling" almost
comically suggests Pound's observation that poetic intuition outpaces more
scientific observation: "Ignorant men of genius are constantly rediscov-
ering 'laws' of art which the academics had mislaid or hidden" (ABCR,
14). One would have to turn that slightly, for Bunting discloses a lawlessness or "play" (*paidia*) that subsequent academic readers of Pound
have chosen to ignore. While the careful philologist would remark the
grammar of the formula, the literary historian would question the belatedness of his protégé's accidental "discovery." Far along into the history
of his own efforts at poetic concision, then, Pound credits a modern English poet with a German-Italian pun that grounds his American poetics.
Thus he tropes a linguistic accident into method.

If he seems indifferent to the irony of elaborating a metaphysical definition ("idea of poetry") on an etymological error already doubly translated, later in *ABC of Reading* he invokes etymology in order to claim
antiquity for his neologism:

DICHTEN=CONDENSARE

This chapter heading is Mr. Bunting's discovery and his prime contribution
to contemporary criticism, but the idea is far from new. It is as we have
said *ingrained* in the very language of Germany and it has magnificently
FUNCTIONED, brilliantly functioned. (ABCR, 92)

Pound's rather cryptic remark leaves open the question of whether he
could have known how important a function *Dichten* serves in Nietzschean philosophy. Probably not, but in passing from the German verb
(*Dichten*) to the English noun (poetry), his text again intersects philosophy. Guided by Nietzsche, Heidegger, and the readers of these two philosophers, we have come to recognize that the subtlest quandaries over
the relationship between philosophy and poetry, and over the rhetoricity
that intervenes in translations, are abbreviated in *Dichten,* a word particularly difficult for English readers of Heidegger.

For instance, the "Translator's Introduction" to *Poetry, Language,
Thought,* a collection of Heidegger's late "poetic" or mystical essays,
clarifies Pound's quibbling over *Dichten*:

> I had tried "poetize" but it had the wrong connotations and excites an-
> noyance in those who feel for the language, suggesting affectation. . . .
> So it gets translated rather as poetry, or the writing of poetry, and often,
> when the word "poetry" appears it is well to remember its sense as a verb.
> (PLT, xii)

This simple matter of word choice is crucial to Heidegger's desire at once
to privilege poetry's verbal force over philosophical abstractions and to
equate the two in Being (*Dasein*), as an originary force prior to language.[28]

Commenting on Nietzsche's translation of "Anaximander's Fragment,"
thought to be the oldest example of Western philosophy, Heidegger says:

> Thinking [*Denken*] is poetizing [*Dichten*], and indeed more than one kind
> of poetizing, more than poetry and song. . . . Thinking is primordial to
> poetry, prior to all poesy, but also prior to the poetics of art, since art
> shapes its work within the realm of language. . . . *The poetizing essence
> of thinking preserves the sway of the truth of Being.* Because it poetizes
> as it thinks, the translation which wishes to let the oldest fragment of think-
> ing itself speak necessarily appears violent. (EGT, 19; emphasis added)

The priority of thought to language, however "poetic" that thought might
be, places Heidegger in the onto-theo-logical tradition he sought to escape. Again, he has elected Being over the force of language—though
he attempts elsewhere to resist the conceptualizing of this Being. In passing, we should note that metaphor—a metaphor of metaphor—is never-

theless inscribed in his text: to speak of "the sway of truth" is to write of metaphor *more metaphorico,* to be taken off course from Truth by the "movement" of metaphor Derrida analyzes. And, because his language thus fails him, the translator's choice of "poetize," which can also mean to write nonsense rhymes or doggerel, seems appropriate.

Let us relate this back to Pound's text, which seems similarly to posit poetry as a unifying force, if not the Truth before and between languages. Throughout his reading text, Pound emphasizes the incomplete, unequal, and plural state of *languages* and *poetries.* Our earlier example will serve here: "The sum of human wisdom is not contained in any one language, and no single language is CAPABLE of expressing all forms and degrees of human comprehension" (ABCR, 34), to which can be added the "translative effect" that "charges" and disrupts tradition: "English literature lives on translation; every new exuberance, every new heave is stimulated by translation, every allegedly great age is an age of translation" (LE, 34–35). As we have seen with regard to the Chinese ideograph and the later reading formulas, Pound fixes on problems of translation and interpretation in such a way as to increase the instability of language and the number of possible readings. Nevertheless, near the end of "How to Read," in a section claiming that one can "read" or interpret a whole culture by knowing a few fragments of verse, he says that one can read poems without knowing the language in which they are written, throwing the reading process open to the force of translation and simple misprision:

> One does not need to learn a whole language in order to understand some one or some dozen poems. It is often enough to understand thoroughly the poem, and every one of the few dozen or few hundred words that compose it.
>
> This is what we start to do as small children when we memorize some lyric of Goethe or Heine. Incidentally, *this process leaves us for life with a measuring rod* (a) *for a certain type of lyric,* (b) *for the German language,* so that . . . we never wholly forget the feel of the language. (LE, 37; emphasis added)

Thus, incorporating the traces of his favorite scientific jargon into an intuitional poetics, Pound simply takes for granted what he had earlier challenged: that language is a fixed system of equivalents, and that poems are fully and easily translated.

The fact that Pound did not explore the impossibility of resolving the problematic of language he raised in no way diminishes the disseminative force of such contradictions. Nor does it prevent his notion of reading from slipping into various neo-Nietzscheanisms. Having considered how he seems to raise some of the rhetorical questions—the questions of rhetoric as figuration—that Derrida treats, we might note that he also echoes the Heideggerian nostalgia for poetic thinking: "Good writing is coterminous with the writer's thought, it has the form of the thought, the form of the way the man feels his thought" (ABCR, 113).

Pound's search for the virtually oxymoronic "American interpretive literature," which is at once a product of translation, transmission, transformation, and an essence prior to language, places him both on the side of the Nietzschean affirmation of rhetoricity and on the side of the metaphysical nostalgia for a primordial and originary "poetic thought." While he could not have been fully aware of this untenable position (especially because it has been recognized only recently by the current generation of neo-Nietzscheans), he nevertheless repeatedly insists on the terms of its impossibility. For example, he locates phanopoeia, or the play of images, in (a reading of) the Chinese written character, *illogically* yoking the scientific and poetic in natural language, which the metaphysical tradition has called *logos:*

> Fenollosa attacks logic in favour of science. The logic appears to him occidental and the scientific approach to knowledge appears to him to be also the poetic and to be the way inherent in the chinese ideograph as distinct from occidental phonetic writing.
>
> His concrete instance is that if you ask any European: what is red? he will say "Red is a colour."
> What is a colour? Colour is a refraction of light.
> What is light? Light is a vibration.
> What is vibration? Vibration is a form of energy.
> What is energy? Energy is a mode of being.
> What is being? etc.
> That is to say the European mind moves from the concrete known to the general and to the still more general unknown.[29]

Pound's efforts to reverse the effects of abstract language and thought, to unite Agassiz's doctrine of direct observation and Fenollosa's notion of interpretive verbs, frequently composes this sort of illegitimate hybrid out of the various arts and sciences. In "Draft of Canto CXIII," he hopes "Yet to walk with Mozart, Agassiz, and Linneas/'neath overhanging air under sun-beat" (CXIII: 786). Yet such mixtures are as disturbing to science, which has always allied itself with logic, as they are to the modern literary tradition which subscribes to the notion that Poetry is an exclusive and privileged kind of discourse, a language at once before and beyond ordinary language. And it is in his arguments against Eliot's "Tradition" that the methodological implications of his "reading" are clearest and most disturbing.

CHAPTER THREE

Tradition and Heresy
"Dissociation" in Eliot and Pound

> And if you will say that this tale teaches . . .
> a lesson, or that the Reverend Eliot
> has found a more natural language . . . you who think
>   you will
> get through hell in a hurry
>
> Canto XLVI: 231

> If the reader is already at the stage where he can maintain at once
> the two propositions, "Pound is merely a scholar" and "Pound is
> merely a yellow journalist," or the other two propositions, "Pound
> is merely a technician" and "Pound is a prophet of chaos," then there
> is very little hope. But there are readers of poetry who have not yet
> reached this hypertrophy of the logical faculty.
>
> ELIOT, 1917[1]

If in 1917 it took a "hypertrophy of the logical faculty" to define the value and genre of Pound's works, by the 1930s one had altogether to suspend canonical judgment and categories to place Pound in literary history. In 1933, Eliot himself spoke the contradictions he had earlier dismissed in that passage from "EP: His Metric and Poetry" (1917), when, in *After Strange Gods,* he acknowledged that Pound was both "the most important living poet in our language" (ASG, 45) and one of the more dangerous heretics against the whole of Western culture. In the latter book, Eliot concerned himself "not with the aberrations of writers, but with those of readers and critics" (ASG, 24). It was Pound's "readings" in the tradition and his reading procedure that most distressed Eliot. During this time—from, say, 1930 to 1935—the two poets maintained close personal and business relations, but their letters and other works reveal that they hardly spoke the same language and clearly were not writing the same style or genre of literary criticism. Eliot, fast becoming "the dean of English essayists" (as Pound dubbed him in 1934), was a respected authority on English literary matters and a spokesman for the moral values represented therein.

The 1930s marked the beginning of Eliot's efforts to institutionalize his critical doctrine and (his version of) the literary canon. To this end, he published several volumes comprised of lectures he had delivered at universities, classics societies, and other literary academies. The most important of these are *The Use of Poetry and the Use of Criticism* (1933), which separates the critical and creative discourses to define the limits of literature, and *After Strange Gods* (1933), a practical application of the limiting categories of "tradition" and "orthodoxy" to modern literature in general and Pound's Cantos in particular. To these books Pound responded with several reviews and numerous letters. But these hardly changed Eliot's mind. Eliot would go on to refine his conservative theology and classicism in such later books as *Notes Toward the Definition of Culture* (1948) and *Idea of a Christian Society* (1939). All these books privilege the English literary canon, especially Donne and the other Metaphysical Poets, as well as later traditionalists like Dryden, Coleridge, and Arnold, who carried over the classics into a kind of Christian devotional verse that Eliot viewed as originary. Eliot saw himself as part of the tradition he founded, not only as the "individual talent" of his earlier formulation of modernism ("Tradition and the Individual Talent," 1919) but also as its representative man of letters.

At the same time, and principally through Eliot's good offices, Pound was writing (in) a heterogeneous counter-tradition which at once challenged the conventional separation of poetry from criticism and the priority of the one over the other. He polemically refuted the notion of the superiority of English to American literature, Western to Eastern thought, and logical argument to evocative fragments. In the 1930s, some of the reading notes and metaphors he called variously "The New Method of Scholarship," "Dissociations," "Paideuma," and "Kulchur" were gathered into several anthologies and books on general culture: *How to Read* (1931), *ABC of Economics* (1933), *Make It New* (1934), *ABC of Reading* (1934), and *Jefferson and/or Mussolini* (1935). This chapter, while touching on these, will examine other fragments which failed to be assembled into those "guides" and "textbooks." I will show that Pound's approach was far from straightforward, though he still preferred such pedagogical titles as *How to Read* and *ABC of Reading*. These texts, along with their fragmentary advertisements and defenses, can be read in such a way as to pose radical questions of the canon Eliot (re-)presented and the criticism he practiced. Indeed, Pound forced his old colleague back to "the basics" that ground—but also subvert—the interpretations and methodology of literary history. The two writers corresponded about the proper reading of each other's works and of the "Great Books" of the tradition. Sometimes indirectly, they carried on debates in journals, prefaces, notes, and other ephemeral texts. But these disagreements are now largely forgotten.

In large part, Eliot, whose own critical preferences came to determine what was published and how that was read, is responsible for dismissing both his disagreement with Pound and Pound's asystematic, unmanageable late criticism. Notwithstanding readjustments of the modern American canon, which have brought Pound a prominence nearly equal to Eliot's, one simply does not have the whole of Pound's critical text immediately to hand—if such fragments can be termed a whole. Pound wrote for journals out of the literary mainstream and frequently concerned himself with questionable topics impossible to disentangle from his politics and economics. Most of his journalistic remarks are unavailable because they were published in generalist rather than literary magazines; other notes and fragments remain unpublished in various Pound archives. These fragments have been neglected by canonical readers, who usually treat Eliot and Pound as allies in the project—or complementary projects—of defining and legitimizing a modern canon, making that canon a punctuation of the history of Western poetics, Eliot's tradition.

There persists a curious disparity between "original" or "primary" documents and historical documentation of the various phases of the Pound-Eliot relationship. The early years, from 1914 to 1923, have been the subject of many histories of (the birth of) Anglo-American modernism (out of the spirit of Western culture). Yet almost no correspondence remains from this period. Pound lost or destroyed most of the letters he received before 1919; along with these, most copies of the letters he sent the Eliot family disappeared into closed files at Harvard University. And, while Eliot gave many of his own early letters to Harvard's Eliot Collection, the largest number remain unavailable and are likely to remain so—at least until after they are published in Valerie Eliot's longtime forthcoming edition of the correspondence.[2] Moreover, in the early years, Pound was so engaged by his various promotional schemes on behalf of modern artists, poets, and musicians that he tended to avoid commenting on any of the critical differences he might have had with Eliot and his other allies and protégés.

Despite relatively sparse evidence from the participants, a good deal of Pound criticism has concerned itself with the relationship and overlapping interests of the two principal "Men of 1914." One thinks of Noel Stock's and Charles Norman's biographies of Pound, which detail the shared goals and individual practices of the two Americans who variously conquered or inculcated European modernism. And in "Eeldrop and Appleplex" (1917), Eliot (Eeldrop) parodies his relationship with Pound (Appleplex) as an update of the quixotic scholarly adventures of Bouvard and Pecuchet. He seems to predict the randomness of *The Cantos* and Pound's various "ABC's" while he alludes to his own urbanity and reserve: "Appleplex, who had the gift of an extraordinary address with the lower classes of both sexes, questioned the onlookers, and usually extracted full and inconsistent histories . . . entered the results of his inquiries into large note-books, filed according to the nature of the case, from A (adultery) to Y (yeggmen). Eeldrop smoked reflectively."[3] The following addendum suggests the lines of inquiry that would later diverge into Eliot's Possum-like theological orthodoxy and Pound's heterodoxy: "It may be added that Eeldrop was a sceptic, with a taste for mysticism, and Appleplex a materialist with a leaning toward scepticism; that Eeldrop was learned in theology, and that Appleplex studied the physical and biological sciences."[4] In the "London Years," when neither writer was anxious to set rules for the other, such methodological differences were more amusing than disturbing. Referring at once to his scholarship and his experimentalism, Pound praised Eliot as "the only American I know of who has made what I call adequate preparation for writing. He has actually trained himself *and* modernized himself *on his own*. The rest of the *promising young* have done one or the other but never both (most of the swine have done neither)" (letter to Harriet Monroe, 30 September 1914; L, 30).

Eliot's little fiction and Pound's advertisement support the widely held literary historical *fact* that the two writers early on adopted the complementary personae of serious Anglo-American *literatus* (Eliot) and American iconoclast or, in Pound's phrase, "conceited fool."[5] Other reminiscences confirm their agreeable disagreements and slightly differing emphases. Nor is there a dearth of interpretations—both contemporaneous and subsequent—of the milieu, even the "vortex," created by the extended Pound-Eliot circle (which included Wyndham Lewis, Richard Aldington, H. D., Henri Gaudier-Brzeska, T. E. Hulme, Ford Madox Hueffer, and James Joyce). Pound examines some of these artistic currents in *Gaudier-Brzeska: A Memoir* (1916), a reading of the "sculptural tradition" from Paleolithic to avant-garde literature—already a heretical reading across several traditions. Excellent, and even creative, histories like Hugh Kenner's *The Pound Era* (1971) and Herbert Schneidau's *Ezra Pound: The Image and the Real* (1976) have recorded the interplay of art and ideas that stimulated Pound's research into the arcana of several continents and the innovations of contemporary arts and science. Eliot necessarily takes a subsidiary role in books about Pound, but most of these treat the two American poets as allies in accommodating modernism into the tradition.

If Pound went farther in both directions (toward archaism and avant-gardism), Eliot was one of *les jeunes* who profited both from his "connections" and his "readings-around." The publication of "The Love Song of J. Alfred Prufrock" (1917), which Pound accomplished as "Foreign Correspondent" of Harriet Monroe's American magazine, *Poetry,* culminates the London years of their relationship. As early as 1912, also in a letter to Monroe, Pound characterized such (American) translations of European culture into the United States as prologue to "our American Risorgimento," a phrase one associates with his 1930s work on Jefferson, Adams, and Mussolini (L, 10).[6] A rereading of the early years of the Pound-Eliot relationship should be deferred until a clearer dissociation of the two writers—as "readers"—can be made from the documents of the later years. This is not to say that such thoroughness will stabilize the context or finally establish the text of their differences—at least not to the satisfaction of those historians who have written on these topics. Indeed, one finds that at its beginnings modernism is a play of revisions, including the literary historical revisionisms of Eliot and Pound.

Perhaps the most important Pound-Eliot exchange takes place over *The Waste Land* (revised by Pound in 1921 and finally published in 1922). Valerie Eliot's facsimile of the correction copy of that poem and the relevant letters suggests that Pound was more than a helpful editor. In Eliot's much quoted phrase, he was *"il miglior fabbro,"* Dante's name for Arnaut Daniel, and part of that earlier Renaissance that Pound and Eliot hoped to repeat—however differently. If only because it produced that poem which represents (American) modernism, the collaboration over *The Waste Land* is crucial, more than two poets reading and writing together. Indeed, it is a chiasmatic crossing still able to generate meanings and reversals of the poets' relationship and the epoch it has come to exemplify. *The Waste Land* made Eliot one among many precursors to *The Cantos,* while Pound was the progenitor and publicist of Eliot's poem. In a suppressed dedicatory "poem," Pound characterized the complexities of their exchange in sexual metaphors which undo clear influence even as they emphasize the ambiguous sexuality of the reading act:

> These are the poems of Eliot
> By the Uranian Muse begot;
> A Man their Mother was
> A Muse their Sire.

How did the printed Infancies result
From Nuptials thus doubly difficult?
....
Ezra performed the Caesarean Operation.

                                (L, 170)

This exchange of roles has been treated in Ronald Bush's *The Genesis of Ezra Pound's Cantos* (1976); Donald Gallup's bibliographic essay, *Ezra Pound and T. S. Eliot: Collaborators in Letters* (1970), takes *The Waste Land* as the epitome of their relationship. With the importance of this exchange one cannot quarrel. But if their disagreement does not mar Eliot's poem, Pound's letters suggest why his many interests are anathema to Eliot's growing religious and literary orthodoxy. For example, the same dedicatory verse advertises Pound's readings in Gourmont's biologism and his orthographic investigations:

> He writes of A.B.C's
> And flaxseed poultices,
> Observing fate's hard decrees
> Sans satisfaction;
> Breeding of animals,
> Humans and cannibals.

                                (L, 170)

This chapter will show that the letters, journalistic debates, and other documents of the 1930s exchange, which inscribe the greatest divergence of the interests of Pound and Eliot, are equally productive of a modernist poetic, perhaps a "postmodernist" poetic, as were the earlier alliances of modern poetry. They yield not only a great modern poem but also a series of questions about such categories as "modernism," "poetry," and "criticism," all nascent in Pound's earlier collaborations with Eliot.

In the 1930s Eliot was Pound's editor. And Pound's replies to Eliot's criticisms, his readings of Eliot's readings, show that Eliot was *behind* Pound's reading text—as its target, if not as unwitting author of the metaphors and other verbal games Pound played upon his friend's more conventional criticism. In this way, Pound's readings—and his notion of reading—are articulated not only in reviews which themselves make little sense except as marginalia to Eliot's books but also in those fragments which remain tangential to Pound's own corpus. These notes, many of which might at first reading seem only marginal to the Eliot-Pound relationship, suggest that Pound frequently deployed his reading procedure against Eliot's attempts to order *The Cantos* and correct Pound's (other) "aberrant readings." More importantly, they insist on his idiosyncratic readings and the interpretive function of poetry and all language. Pound's strategy, at least in part, is to evade Eliot and the conventions of orthodox critical books; in this, at least, Pound's readings succeed. They remain outside his corpus, outside even the posthumous books, and will probably never be as well received as his earlier criticism.

No books have been written about his later criticism as such, though several analyze the tangle of notions comprising his Fascism and idiosyncratic economics. Some of these gloss the difficult late Cantos, while others reconstruct (or apologize for) the political climate in Pound's Italy. Books like C. David Heymann's *Ezra Pound: The Last Rower; A Political Profile* (1976) and Earle Davis's *Visions Fugitive: Ezra Pound and Economics* (1968) provide useful biographical and ideological contexts for Pound's apparent misprisions. Michael Bernstein's *Tale of the Tribe: Ezra Pound and the Modern Verse Epic* (1980), an account of Pound's founding efforts on behalf of an epic tradition in "the American idiom," which incorporates contemporary politics, economics, and history, treats the late prose as part of this synthesis but finds *The Cantos* a great poem despite, and apart from, the didactic errors of the prose. None of these books rigorously addresses the Pound-Eliot relationship of the 1930s. One of the few which treats this difficult late period as anything more than an addendum to the earlier collaborations is Timothy Materer's *The Vortex: Pound, Eliot, and Lewis* (1979). As its title suggests, this book is focused on 1914 and Pound's Imagist/Vorticist criticism as the genesis of modernism; it addresses *After Strange Gods* and Eliot's other corrections of Pound, but with so pronounced an Eliotic bias as to dismiss the later Pound with Wyndham Lewis's famous epithet, "the revolutionary simpleton."

But it is not my principal business at this point to speculate why critics have virtually ignored the marginal texts I wish here to uncover. Still, Pound's reading books, like his fragmentary journalism, are never far removed from his political and pedagogical commitments to revise American and all modern culture. Since orderly or logical critical discourse and the formation of the literary canon(s) are among the fundamental notions and methods interrogated by Pound's reading procedure, one should perhaps resist dismissing these fragments simply because of the (quite considerable) difficulty of placing them in Pound's corpus or within the modernism canonized under his name. Moreover, one of Pound's objectives in his reading texts, in all his writings for that matter, is the breaking up of the discursive categories which separate the various arts and sciences into mutually exclusive and strictly hierarchized disciplines. The heterogeneous texts this chapter will touch upon—and allow itself to be touched by—imply radical questions of generic and historical demarcations when they do not ask them outright.

II

The majority of those who have used Pound's later criticism for anything more than a gloss of his poetry or as a literal explanation of his Imagist/Vorticist *theory* have tended to devalue if not disparage it, to reduce it to compendia of citable aphorisms and quirky formulations. One can hardly deny that his notes on reading fail to yield a system or even a stable series of master figures. And this seeming chaos of theory, this "lack of high seriousness," if you will, has disturbed those few serious critics who have treated Pound's reading in any systematic way. In this regard, F. R. Leavis's response to *How to Read* is typical. While nodding to Pound's erudition and his stimulation of thought, he protests the eclecticism, pluralism, and heterogeneity of Pound's artistic models. Virtually echoing Eliot's demurrers in *After Strange Gods,* Leavis, in a pamphlet entitled *How to Teach Reading: A Primer for Ezra Pound,* takes Pound's eccentric pedagogical stance at its face value and offers his own "Parson's" corrective to Pound's cultural deviations. Leavis attacks Pound's failure to understand the relationship of literature to language, which in Eliot's terms has to be regulated by a traditional Mind, rather than a democratic or individual one. But if Leavis fails to appreciate the radical potential of Pound's interrogation of the reader's conscious control and memory, he does indicate what is at stake in the debate between Eliot and Pound when he says of the latter:

> Literature for him is a matter mainly of individual works, Chinese, English, Provençal, Anglo-Saxon, and so on, written by individual artists who invent—or borrow from other individual artists—devices, processes and modes of charging language with meaning. . . . Not only is language an apt analogy for literary tradition; one might say that such a tradition is largely a development of the language it belongs to if one did not want to say at the same time that the language is largely a product of this tradition. Perhaps the best analogy is that used by Mr. Eliot in *Tradition and the Individual Talent* when he speaks of "the Mind of Europe." "Mind" implies both consciousness and memory, and a literary tradition is both: it is the consciousness and memory of the people or the cultural tradition in which it has developed.[7]

In what is only the earliest in a long series of Eliotic supplements to Pound's fragmentary reading text, Leavis treats both writers as traditional readers and assumes that in *How to Read* Pound merely fails to preserve intact the received tradition of canonical texts to teach a hermeneutical method for extracting the proper meanings from the text of literary history. Therefore, Leavis repairs Pound's apparent disorganization and omissions by substituting for them Eliot's more coherent readings. He also assumes, then, that Pound subscribed to an organic model of language grounded in the privilege of cultural consciousness. If Pound's loose and changing definition of "Kulchur" as uncontainable and virtually unconscious, beginning "when one has forgotten what book" (GK, 134), was unavailable to Leavis (it was published some six years later), his fusion of Pound's heterogeneous readings and Eliot's tradition seems to ignore the disruptive potential which Eliot himself acknowledges.

One should recall that Eliot branded Pound a heretic for such (mis-)readings in the tradition. Pound was not the only writer accused of the aesthetic "heresies" of exoticism, secularism, and individualism. *After Strange Gods* also indicts D. H. Lawrence's psychologisms, Yeats's mysticism, and Irving Babbitt's Confucianism. Yet Eliot found *The Cantos'* use of Dante's theology and poetry particularly disturbing, for this not only involved the secularism he wanted to halt but also disturbed the very tradition he invoked against such heterodoxy. It was primarily at Pound—who was attempting to revalue tradition, not merely "to value himself too highly" as innovator or prophet of a new religion or system—that Eliot directed these words, which virtually describe Pound as one of the "tempters" in *Murder in the Cathedral:*

> It is characteristic of the more interesting heretics, in the context in which I use the term, that *they have an exceptionally acute perception, or profound insight, of some part of the truth;* an insight more important often than the inferences of those who are aware of more but less acutely aware of anything. *So far as we are able to redress the balance, effect the compensation, ourselves, we may find such authors of the greatest value. If we value them as they value themselves we shall go astray.* (ASG, 26; emphasis added)

Eliot does not dispute Pound's claim to the heterogeneous details or partial truths of the literary tradition; indeed, these are the very things he hopes to balance or "save" for the system of literary valuation he maintains in *After Strange Gods*. The theological metaphor is strategic, if not in precisely the same way as Nietzsche's similar troping of Wagner. One comes to recognize that book as a defense against a whole series of aberrations that Pound dispersed in various letters, poetic fragments, and journalistic pieces that occasioned and followed Eliot's corrections. These marginalia, which Eliot himself took more seriously than the formal reviews of *After Strange Gods,* illustrate the many-sided attack Pound launched against Eliot's critical and editorial program.

Hardly limiting himself to critiques of *After Strange Gods* and *The Use of Poetry and the Use of Criticism,* Pound articulated his own ongoing reading procedure in the interstices—if not the very pages—of *The Cri-*

terion, various Faber & Faber publications, and the generalized "establishment press" he made Eliot represent. Therefore, before turning to the text of the debate over *After Strange Gods* proper, one should note its context, its pretext, and its pre-texts. The mergings among this randomly articulated creed repeat the instability Pound affirmed in the canonical readings he there attacked.

As previously suggested with regard to Eliot's books, in the 1930s these two representative American modernists were as deeply engaged in criticism as they were in poetry—insofar, one might add, as it is possible to separate these two discourses in Pound's work of this period. In Rapallo, Pound was writing *The Cantos,* especially the "Jefferson/Adams" unit of *Eleven New Cantos* (Faber, 1934), which attempts to replace the vitiated and borrowed English belletristic tradition with the "monument of the Jefferson-Adams letters," thus privileging political conversations over poetry and American revolutions against European civilization over any notion of continuity with literary classics. He was also working on "The Usury Cantos" of "The Fifth Decad" of *The Cantos* (Faber, 1937). Published journalistically, these elicited Eliot's most serious strictures against the politicization of Dante and the other poets who provided Pound a *theory* of usury. The usury argument informs *After Strange Gods* and its reviews. At this same time, Pound and Eliot also corresponded extensively over revisions of *A Draft of XXX Cantos* (Paris, 1930; Faber, 1933), that poem which so infuriated classicists and translators. Eliot edited and published all these as one of the directors of Faber & Faber, which was fast becoming the most prestigious modern press—but not, it should be added, on account of the sale of Pound's relatively unpopular poetry and his virtually unsold critical books.

Pound did not limit his activities to writing and revising poetry. Most of his time was spent writing Social Credit propaganda and propagating his own literary theories—granted "theories" is always a problematic term for Pound's critical notions, which fall somewhere between practical if generalized readings and prolegomena for future methods and theories. The status of his critical text is particularly difficult to establish with regard to the textbooks on reading; these inscribe a reading *procedure* which implies program, method, even ideology, but refuses systematization. Continually mixing method with metaphors, he wanted to prescribe "the right hierarchy" and yet play the anarchist. In one of several letters calling upon Eliot to use *The Criterion* to promote the pedagogical aesthetics of *ABC of Reading,* which Faber & Faber had published at nearly the same time they had rejected *Jefferson and/or Mussolini,* the text marking his final break with any tradition defined by the literati, Pound both derides

Eliot's respectability and hopes to exploit it for a return to his own preferred work: "I don't see that the circulation wd/now be endangered by
an assertion of the true hierarchy of values. . . . I don't see who can
DO the job of pouring the acid if you refuse. I ought to be left free both
for the Cantos and Econ."[8] Eliot was of a mind to do anything but pour
acid, one of Pound's recurrent metaphors for the "antiseptic" and transformative—even deformative—functions he affirmed for poetry and criticism, which were to keep thought and language active. As Pound never
tired of reminding his former literary cohort, Eliot was too involved with
the London literary establishment to circulate the "right" (Poundian) economic and related aesthetic notions.

Because he could not rely on Eliot for an astringent style, let alone
substantive attacks on the literary establishment, Pound adopted his own
strategies, which is not to say that he wrote in the (Eliotic) style or mode
of argumentation fast becoming standard. Indeed, not even the *ABC of
Reading,* advertised as a toned-down version of the "spiky parts of the
author's earlier critical skirmishing" of *How to Read,* and thus "imper-
sonal enough to serve as a text-book" (ABCR, 11), ever ceases "pouring
acid." Like the insistently fragmentary notes and poetic fragments of concern to us here, that book refuses to argue a single thesis through a series
of examples. And what is more disturbing, its loosely assembled fragments spill over into his other writings. Pound's various reflections on
reading, as we saw in Chapter Two, provide neither exhaustive readings
of the texts they canonize nor a method to master such texts. Instead, at
every turn, they seem to subvert the critical enterprise rather than to advance a serious new criticism. Yet this is in keeping with Pound's ambivalence about the task of critical reading, his conflicting desires to master poetry in exhaustive or comprehensive analytics, and yet to allow poetry
its full disseminative (and paradoxically "totalitarian") power to exceed
all critical reductions. In these texts, and throughout his criticism, he
questions the relative value of literature and criticism, writing and reading, poetry and prose.

On the one hand, he can say that "poetry is totalitarian in any con-
frontation with prose. . . . Certain things are SAID only in verse. You
can't translate 'em" (GK, 121). On the other, his translations back and
forth among various languages and traditions overturn the privilege afforded poetry and raise such discourses as propaganda and criticism to
equality with, even superiority over, the classics Eliot canonized in the
purely literary "tradition." Thus, in an unpublished table of contents for
a proposed volume of his collected prose, Pound says: "There is no valid
reason why the re/recation [sic; but is the correction "reaction" or "re-

creation"?] of beauty should always fall below the original. The supposition that it does is half the time but the fruit of a complex of inferiority in the sterile."[9] Through all his works one seeks in vain Pound's final judgment of the value of criticism and that of his own critical text. At times, his metaphorics suggest a generative, even disseminative, function for criticism—anything but "sterile" academicism. Elsewhere, inviting chaos, he grants the reader, his readers, a certain transformative power: " 'It is not what a man says, but that part of it which his auditor considers important, that measures the quantity [sic?] of his communication' " (GK, 59 [attributed to Frobenius]).

Rather than final insights into Pound's "intentions" or "authorized readings," his fragments concerning reading gather together the self-reflections on genre and value endemic to Pound's text: what belongs to *The Cantos* proper? What to his reading texts? Instances of his mixing, reinscribing, and revaluing fragments from letters, essays, poems, music, and other heterogeneous documents (his own and others') into hybrid commentaries and/or poems are legion. In many of the spaces between his greatest poem and the negligible unpublished fragments, one discovers Eliot and their argument over proper readings. While I have for the most part limited myself to writings from the period of *After Strange Gods,* I might cite the poignant lines he wrote Eliot from Yale when, in 1959, he was organizing these fragments into the folders somewhat inappropriately titled "Collected Prose": "Sitting in my ruins and heaven comes down like net (and all my past follies)."[10] These lines are woven back into the "Draft of Canto CXVI": "a tangle of works unfinished. . . . I cannot make it cohere. . . . i.e. it coheres all right. . . . And as to who will copy this palimpsest?" (CXVI: 795–97). Taken together, these describe the paradoxically elliptic and redundant notes that comprise Pound's reading texts and, for the most part, the kind of critical polemics he engaged in the 1930s.

Curiously, many of the unpublished fragments provide the explanations and summaries absent from the published criticism. If they cannot be called programmatic statements, they nevertheless gloss the more well-known reading texts and indicate the importance of the Eliot-Pound relationship to these. One such fragment, written in defense of the style of *ABC of Reading,* sarcastically indicts Eliot's sort of corrective or prescriptive reading, but, at the same time, it calls for the "intelligent," "sane," educated, if not academic, reader Eliot was training:

> Then there is *Mr. Q.T.* who finds prose rough when it doesn't put him to sleep, still *slumbering over his grammar book and doesn't know that the sane and intelligent reader will welcome an ellipsis* that saves 40 words

*even if the ellipsis jump 2 or 3 clauses and leaves what looks like a gram-
matical error.*[11]

Ellipses, whether or not the trope is signaled by the punctuation mark of
the same name, characterize Pound's reading notes. Sometimes these do
cause ungrammatical sentences; more often than not, they contribute to
his strange reading economy which generates pages of commentary out
of the gaps left by a few "saved words." Just so, one easily becomes
distracted by the private conversations, the unmanageable "background
noise" of that passage. Perhaps Eliot, who as his editor at Faber & Faber,
had occasion to correct Pound's prose, is implicated in "Mr. Q.T." "QT"
is slang for quiet, on the sly, by extension, the self-effacing (Eliotic)
reader—quiet as an "Ol' Possum." Elsewhere Pound names this proper
reader, the writer of correct essays, "Eliot." For example, to Eliot's attempts to correct the essays assembled in *Make It New* (1934), he re-
torted: "Any fahrt can write orderly essays."[12] There seems, then, a certain method to the inter- (even intra-)textual madness of his reading which
can be said to inhabit the readings it motivates.

Pound offered his views on reading in the many letters he wrote concerning the volumes Eliot was editing. This is not to say that his position
is any clearer there. And Eliot felt obliged to comment on his old friend's
strange epistolary habits, often in his "Eeldrop and Appleplex" manner:
"I will arise & go now & go to Rappaloo/ where the ink is mostly green
& pencils mostly Blue."[13] While it might seem off color to base a reading
of their differences on such doggerel, one might note Eliot's concise reminder of the textual curiosities that distinguish Pound's original typescripts and manuscripts. Pound, however, denied any feelings of isolation, and his massive journalistic output would seem to place him in a
literary scene—however dispersed.[14]

For all the letters they exchanged, the two poet-critics hardly maintained a dialogue; rather, they reported to each other the results of their
individual works in other areas. This is especially true of Pound's letters,
which are full of solecisms that required explication in follow-up letters.
Since Eliot had his hands full with correcting Pound's spelling and grammar, as well as his inattention to copyright and the other conventions and
legalities of publishing, his letters are more businesslike. While he exhibits a certain self-consciousness, he still enjoys dismissing Pound's voluminous correspondence. In a letter that failed to dissuade its recipient
from sending more letters, essays, poems, and fragments, he jokingly
describes his Faber & Faber tasks thus:

> the young *enterprisin* firm of *Faber & Faber is forchinite in having* secured
> the services of *Mr. T. S. Eliot. . . . After 12 years with the Federation*

of British Industries Mr. Eliot is probly better qualified than any other Big Executive to cope with the task for which he has been engaged, which is *to superintend the department devoted to correspondence with Mr. Ezra Pound.* (emphasis added)[15]

There, in imitation of Pound's own idiolect, Eliot describes the inflated economy of his communications with Pound. In response to simple corrections and suggestions, Pound would dash off pages of defensive explanations and notes from his economic and political readings. Eliot seems also to comment on Pound's growing enthusiasm for economics. If the coincidence of economics and the economy of Pound's writing seems gratituous in Eliot's puns ("enterprisin," "Executive"), it is confirmed by Pound's letters, which became longer and more numerous as his concern for economics deepened. This is in keeping with the "productive exchange" he affirmed in both poetry and personal correspondence.[16]

Not that Eliot always approved, or even deciphered, Pound's communications. In a letter that might explain his forgiveness of Pound's offensive critiques and his refusal to incorporate some of Pound's explanations into his editions of *The Cantos* and the prose, Eliot says: "It's always a pleasure to receive a letter from you. . . . It's equally a pleasure even when, as is usually the case . . . I am quite unable to crack the cipher."[17] Eliot's friendly bemusement notwithstanding, the two writers were in greater disagreement than either was willing to admit—perhaps than either recognized. Their ideological differences can be read as much in their conflicting expectations and interpretations of their own works in progress as in their explicit arguments over critical positions. Pound's idiosyncratic notions, his growing alienation from the central ideas and practices of literary London, required that he send Eliot lengthy glosses of *The Cantos.* These readings exemplify his reading procedure which also becomes involved in that poem, thereby creating the sort of "marginal commentary" Pound favored over polite essays. In writing about the *Draft of XXX Cantos,* he rejects Eliot's efforts to keep that poem within proper generic bounds. His refusal to omit the line numbers of his rather obscure translation of *The Odyssey* can be viewed as part of his reading procedure:

> *Gordamn that greek/I putt the line numbers in the margint, thining anybody that cd reet it at tall cd/tell wot bok it cum from,* namely the Circe or Kirke book of the O/dishy. Look at lines 213, 228, 317.
> *I don't quote in full, I leave out wordz that don't comport with wot I'ma driving at.* (emphasis added)[18]

Since he regularly fails to give complete citations, let alone summaries or explanations, his line numbers and allusions frequently lead the reader nowhere—even if there are signposts enroute to an interpretive abyss. In the above example, one is led through such mediating texts as Divus's translation of Homer back to Pound's idiosyncratic version of Homer. Not only is this explanation written in Pound's "Amurikan" or Canto(n)ese; it also violates at least two of the laws of modern (English) letters Eliot tried to enforce: it defies any notion of self-contained, self-reflective, or, in Eliot's word, "autotelic" poetry; moreover, it insists on merging commentaries (on *The Odyssey* and/or on *The Cantos*) in a manner that disturbs the incorporation of modern and traditional poems, blurring the values of literature and criticism. Such a reading—or palimpsest of readings—cannot be called either poetry or criticism. It is nevertheless of a piece with Pound's reading notion and procedure, of interpretative readings which likewise promise and resist totalization. Pound's concluding sentence gives another reason why his readings are unacceptable: "I leave out wordz that don't comport with wot I'ma driving at." While such a principle of selection might be said to underlie all interpretations, it is an embarrassment to serious criticism, not to say the idea of translation that would conserve a stable meaning (signified) through a universal exchange of words (signifiers).

Understandably, one might (mis-)read Pound's rigor, which borders on textual fetishism, as simple error—especially in letters where unconscious typos are indistinguishable from witticism. Here one has recourse to Paul de Man's description of Nietzsche's similar "rigor" which, in the guise of error, at once parodies and exposes serious or philosophical criticism. If one cannot feel confident of Pound's mastery of all the languages he uses in, say, *The Cantos*—all the philosophemes, psychologisms, biologisms, and whatnot he quotes, translates, or otherwise invokes—these nevertheless play havoc with the soundest principles of textual scholarship, bibliography, and hermeneutics. In this way, he affirms the infinite regress of interpretation which most interpreters attempt to halt; at the same time, he calls into question the very idea of the translator's mastery and the translator as subject or regulator of exchange. Thus Pound's readings are not simply "open-ended"; they are open at both ends: the reader cannot fix Pound's meaning, nor has Pound fixed the meaning of the original text and its context in literary history. This is in keeping with his notions of creative translation and the "increment of association" or, in a phrase, his ab-usive reading. Such foul play has not deterred readers from source hunting in *The Cantos,* from assuming that Pound accepted the general rules, the canonical values and procedures of literary scholarship that define allusion, quotation, and so on, by philological laws. As is easily seen in his remarks to Eliot about the basics of textual transmission and authorial integrity, Pound simply did not play by the rules.

In a letter dated 31 January 1934, Eliot simply asks about *Make It New:* "Can you reassure us about all copyright? There is a Hell of a lot of quotes."[19] Notwithstanding his own legendary attempts to enforce copyright restrictions on the various printings and translations of his own works, Pound sent the following manifesto-like response:

> *One of the diseases of contemporary* [?] (*and probably running back 100 years or more*) *is due to the loss of making commentaries. I mean marginal commentary* on important texts.
>
> This is due in part to the conditions of the god damned book trade and part, I regret to say, to the laws of copyright, no benefit without a malefit. (emphasis added)[20]

While he would seem to subordinate commentary to "important texts," suggesting the separation of literature and criticism and the privilege of the former, he instead insists on the coincidence of original text and commentary. Just as he praises the least literal of translators (Golding and other Elizabethans who translated Latin texts of the Greek classics), he prefers readings that point up the distance from, not the proximity to, univocal meaning and original work. Not only does he play against (Eliot's) attempts to stabilize literary categories; he finds precedents for his procedures in traditional texts. If he does not go so far as to announce the destruction—or dissemination—of tradition, he hopes to correct modern practice by invoking a heterogeneous tradition, a catalogue of readings.

In another fragment, Pound notes that book publishers will not accept unadorned or unincorporated commentaries. They want original work or internally coherent critical arguments, not notes that supplement books. In keeping with his strange reading economy, he advocates revisions that undo books, not additional and more exact books. He does not want simply to save words, but also to circulate more words and ideas by rereading:

> *There is no "book to be made" from a collection of errors* yet the rectification of a good book containing say 20 years work of an intelligent but far from omniscient author wd. be more helpful than the making of a new book on more or less the same subject. . . . *Work "partly" done by periodicals,* but *the serious criticism in a review remains forever separate from the original book.* (emphasis added)[21]

There he clearly blurs the distinction between original and commentary, book and *reading*. As he asked "Why Books?" in *ABC of Reading,* Pound affirms the style of juxtaposed quotes, unexplained borrowings, and the other aberrations of popular periodicals. At the same time, he complains that establishment reviewers prevent the productive, if disruptive, merging of texts.

In a review of Silvio Gesell's *Natural Economy,* Pound specifies his objections to journalists who maintain the mutual exclusivity of literature and the *other* discourses. He offers a corrective to the correctors in the form of a tradition that violates the constraints that the nonpoets would place on poetry:

> *The diseased periphery of letters is now howling that literature and poetry* in especial *should keep within bounds . . .* that literature should eschew the major field by omitting and leaving untouched a great deal of the subject matter that interested such diverse writers as Propertius, Dante, and Lope de Vega. (SP, 272; emphasis added)

Notably, he does not challenge the notion of "major field." Nor does he affirm the value of being on the periphery of several disciplines. Instead, while seeming to reject such adulteration, he allows nonliterary subject matter to insinuate itself into the very words used to make the distinctions he abolishes. Typically, he affirms a countertradition in the privileged terms defined by the orthodoxy he attacks. As Eliot will ask, what can "poetry" and "literature" mean in Pound's lists of diverse precursors?

Yet Pound looked back fondly to a time when journals freely published his sort of disruptive, and not purely literary, criticism: "Twenty years ago little magazines served to break a monopoly, to release communication, mainly about letters, from an oppressive control" (SP, 272). Rather than seizing control of a journal in order to institute his reforms, as he had in the earlier years, for example, as foreign editor of *Poetry,* Pound tended to attack the ideas and editorial policies of establishment literary journals from the outside. Therefore, he frequently contributed his notes and essays to little magazines that one simply omits from the category "serious criticism." But his remarks about Gesell were published by Eliot in *The Criterion.* It was that journal and its conservative editorial policy that Pound, thanks to Eliot, could subvert from within.

Pound's attack on the oppressions of literary journals serves as a reminder that, in addition to his Faber & Faber duties, Eliot controlled *The Criterion* from 1923 to 1939. *The Criterion* was a fine paper quarterly, a *journal* in the bookstore or academic sense—not a newspaper, *le journal,* of the type Pound and Remy de Gourmont preferred.[22] It was devoted to substantial, both lengthy and authoritative, essays by established critics, scholars, and philosophers. Most prominent among these were F. R.

Leavis, I. A. Richards, A. L. Rowse, and Jacques Maritain—all names that will resurface in Pound's assaults upon *After Strange Gods*. The essays Eliot published conform to the academic, if not the Oxbridgean, style and remain well within the boundaries of *The Criterion*'s subtitle, *A Literary Review*. While some contributions broached political and economic issues, they did so with an ecumenicalism which, in keeping with Eliot's practice, emphasized literature and philosophy over economics and politics. The choice of the former, to the near exclusion of Pound's latter two enthusiasms, is a recurrent argument between the two writers. Perhaps because Eliot's privileging of literature, or more precisely aesthetics or belles lettres, to the exclusion of the other discourses, seems altogether appropriate for a "literary review," one tends to accept Eliot's reaction against Pound's politics as a justified editorial decision, rather than what it was, an active resistance to Pound's critical argument against the division of labor that would isolate literature and literary criticism from other (impure) genres.

Eliot did include Pound's "Murder by Capital" in *The Criterion* of July 1933. But, as though to apologize for momentarily violating his own strictures against propagandistic and socioeconomic intemperance, he was prompted to clarify his own editorial position. In the "Commentary" of the same issue, Eliot somewhat immoderately insists upon his own moderation. One recalls Pound's comment in another context: "During the past 20 years the chief or average complaint against the almost reverend Eliot has been that he exaggerated his moderations" (NEW V:12, 297). Eliot acknowledges the journalistic climate in which economic, political, and aesthetic concerns seemed again to require *manifestoes* (as they had in 1914, when Wyndham Lewis's *Blast,* the avant-garde little magazine, had published "The Vorticist Manifesto"). But he demurs: "It seems to be the necessity of the moment—at least in America—for the editor of a literary periodical to explain exactly where that periodical stands on the great political and social issues of the day. I have no intention of doing that myself on this occasion; and I have not yet framed any manifesto against manifestoes."[23] Yet he does write such an antimanifesto, one that argues for some manner of system against the chaos of political and economic concerns which interested the likes of Pound:

> We cannot say that the [economic] emergency requires first a readjustment in the political-economic world, and that when that is effected we may turn our attention to making it a world in which there is a positive value. *The system which the intelligent economist discovers or invents must immediately be related to a moral system. (The Criterion* 12: 49, 649; emphasis added)

That necessary and mystical incorporation of other, more worldly systems into the universal moral System echoes and theologizes the resolution and ordering of innovation into the tradition that Eliot repeatedly averred after the 1919 publication of "Tradition and the Individual Talent." More importantly, it sets the priorities, now traditional in American literary politics, of the universal (the spatial writing of Poetry) to particulars (the temporal conflicts of politics); these are also reflected in an elevation of form over theme. Pound was vigorously to attack this nostalgia in his critiques of *After Strange Gods.* He had also done so in his many proposals for magazines that would address the heterogeneous discourse (Poetry) he privileged in Dante and in the ephemeral economic-political-aesthetic journals of the 1930s.

As is well known, throughout his career, Pound's letters are full of schemes for popular journals. Indeed, time allowing, one could trace his increasing alienation from the literary critical mainstream through his many "Proposals," "Prospectuses," and "Manifestoes" for journals he was never to edit.[24] And he made at least one concrete suggestion that bears on Eliot's journalistic policies when, in a letter of January 1935, he proposes that the two of them collaborate in "a new paper/mag that is $\frac{1}{2}$ orthodox Douglas and $\frac{1}{2}$ 'Intellectual Liberty' selected cause 'Alive.'"[25] In contrast to *The Criterion*'s orderly essays and the twenty or so pages of careful reviews of current books included in each of its issues, Pound suggests that "It wd/be nice if you wd/reserve 4 pages per issue, to tell the reader honestly WHAT is fit to read."[26] Perhaps Pound wanted to revive *The Exile,* a quarterly consisting of poems, stories, and unexplained fragments of current affairs. In any case, one might turn to that publication as his ideal of a certain asystematic journalism.

Though Pound had served as "literary adviser" and "foreign correspondent" to several American and transatlantic literary reviews (*Poetry, Little Review, The Dial,* even Orage's social credit papers, *The New Age* and *New Democracy*), *The Exile* was his only venture into publishing and editing his own magazine. Curiously, throughout *The Exile*'s brief run (four quarterly issues, Spring 1927 to Spring 1928), Pound was unusually silent about his critical views. This editorial *praxis,* thin as it is on anything resembling theory and logical argument, is consistent with his notion of reading. One obvious feature—inasmuch as an absence or omission can be said to be obvious—that distinguishes his magazine from serious literary reviews is its almost complete lack of organization and explanation. The apparent randomness of the stories, poems, notes, and fragments in the four issues of the magazine might be thought to exemplify Pound's "presentational method" (NA X:10, 225) or "the ideogrammic method" (CWC) or the biologist's method of comparing slides (ABCR, 22 *et passim*). In any case, if the ideology underlying his magazine remains obscure, Pound insisted that he was publishing those "banned writers" who represented modern thought at its most active. While he had not yet made explicit the connection between reading and general economies, his goal was to facilitate the "exchange of ideas." And, if only because his most frequent contributors (including Robert McAlmon, William Carlos Williams, and Basil Bunting) later wrote for several American Social Credit newspapers, economics and literature can be said to merge in *The Exile*. Both discourses are involved in his effort again to tamper with copyright and some of the other conventions of the book trade.[27]

In lieu of an editorial statement, Pound launched his magazine with a broadside advertisement consisting of a quotation from "Part of Canto XX" and "Adolphe," John Rodker's short story, which appeared in the first number of *The Exile*. Pound used this advertisement to obtain American copyright, but because he had titled the first issue "Spring 1927," the "American booktrade" defined his publication as a book, not a periodical. As such, it would be subject to taxes and not fully protected by copyright. His failure to obtain copyright became the conspiracy against the free circulation of ideas that he reported in various indictments of "The System." It was neither the first nor the last of Pound's battles against genre and nationalism. As though to defy Eliot's and other canonical (or canonizing) definitions, his transatlantic magazine was a translation of (European) *Kulchur* to the American hinterlands—or, since most of his contributors were Americans unread at home, a retranslation of American poetry.[28]

The Exile seems nearly to parody the editorial statements of critical journals. Against programmatic or systematic (Eliotic) readings, Pound affirms the independence of the artist who, as privileged interpreter, is at once in advance of and outside all political and aesthetic revolutions. In characteristic metaphors—more precisely, in a *polyptoton* or translation of the same word into several cases—he undoes grammar and the political metaphor of "revolution": "The artist, the maker, is always too far ahead of any revolution or reaction or counter revolution or counter-re-action for his vote to have any real result" (*The Exile* 1, 88). That translation from politics to aesthetics—or so it must be read in retrospect—is as disturbing to politics as it is to art, since it unsettles the language that would stabilize their separation and define positions within the two discourses. In the same vein, and in another text intended to present "Active Artists" to the stagnant literary establishment, *The Active*

Anthology (Faber, 1934), Pound turns this figure against Eliot: "I note that if I was in any sense the revolution I have been followed by the counter revolution" (AA, 8).

That ideological irresponsibility, an abuse of etymology and an abomination to any ethical use of language, should perhaps not be taken entirely seriously. Still, it enables Pound to canonize writers as incompatible as Mussolini and Lenin. For example, in *The Exile* 3 he quotes the following without benefit of gloss, reference, context or any of the things required to make these translated fragments "readable"—in any sense other than Pound's own, as an infinite regress of associations and guesses about his interpretation of some titulary "original":

> We are tired of government in which there is no person having a hind-name, a front name and an address.
> > —B. Mussolini
>
> The banking business is declared a state monopoly. The interests of the small depositor will be safe guarded.
> > —V. L. Ulinov (Krylenko, Podoisky, and Gorbunov)
>
> The duty of a being is to persevere in its being and even to augment the characteristics which specialize it.
> > —Remy de Gourmont
>
> People are not charming enough.
> > —Le Sieur McAlmon
>
> > > > > *(The Exile* 3, 103)

To read that passage one needs the full baggage of Pound's earlier and later readings of the writers quoted. Here one encounters the irony of which Pound is at once master and victim: "'It is not what a man says, but the part of it which his auditor considers important, that measures the quantity of his communication'" (GK, 59). In Pound's readings such irony increases exponentially; for at every stage on the journey to an original thought, he affirms arbitrariness by selecting aberrant passages—not ones representative of a writer's system, but those which "comport with wot I'ma drivin at" (letter to Eliot, 10 February 1935; quoted above).

In passing, one might speculate that in the above passage the Gourmont quotation provides a reading of the other random fragments. It suggests a sort of "biology of style," the principle of "natural selection" of individual—emphatically individual—traits that distinguish styles of thought, action, and communication among the higher species. Pound would seem to judge the propriety of the names of the writers he quotes by such a principle of natural selection or specialization. This playing with names and privileging of distinction and detail Pound elsewhere associates with

Gourmont's "dissociation," with which we will have more to do in the next chapter. This anthology of quotations, however, indicates that Lenin et al. represent the lowest form of government and specialization, rule by committee, in contrast to rule by the properly identified and responsible individual: B. Mussolini. Whatever the intention of the uncontainable associations those unaccommodated quotations might be thought (strategically) to open in the discourses from which they were drawn, it would seem clear that they mock intentionality and final interpretation and threaten any sort of totalization—whether that of an institutional literary tradition or political totalitarianism.

In *The Exile* Pound never states an editorial policy that can be placed alongside that of *The Criterion*. By omission, then, he seems guilty of the sins of which Eliot and Leavis accused him: presenting heterogeneous selections from various traditions and languages, contributing to the incoherence of the modern "mind of Europe." To conclude the present excursus into the journalistic hinterlands, one might note that, as though to counter *The Criterion*'s censure of journalistic excesses, *The Exile* 4 gives an annotated bibliography of Pound's favorite little magazines in the format of "4 pages per issue, to tell the reader honestly WHAT is fit to read." He lists those journals in which he introduced an earlier generation of modern writers: *The New Age, The Little Review, The Dial, The Egoist, Poetry, The Transatlantic Review,* and *The New Freewoman.* To this nostalgia for the heterogeneous past of modernism, Pound appends a list of his own earlier works and prospectuses of works in progress; among the latter are the reading programs and economic tracts of the 1930s. In the final pages of the last issue of *The Exile,* Pound underscores his function as popularizer or avant-garde reader of avant-garde writers. This is, if such a definition were possible, a definition of the aberrant individualism Eliot abhorred:

> It is of inestimable value that there be *men who receive and transmit things in a modality different from one's own;* who correlate things one wd. not oneself have correlated. *The richness of any given period depends largely on the number and strength of such men.* (*The Exile* 4, 107)

Thus, Pound, who both *received* and *transmitted* (his recurrent metaphors for reading) differently from Eliot, would seem to affirm the privilege of those writers and readers who disturb the English belletristic tradition.

Indeed such an emphasis on discontinuity and discrimination in literature is consistent with even Pound's earliest criticism. One might recall the "method" and the similar electrical metaphors of "The New Method of Scholarship," the electrical field theory, as it were, which was pre-

sented in fragments under the title "I Gather the Limbs of Osiris," from
1911 to 1912 in A. R. Orage's *The New Age,* one of the ephemeral journals Pound wanted to canonize alongside the recognized monuments of
modernism. These early essays, which are here quoted from the little
magazine rather than from *Selected Prose,* presented "explications and
translations in illustration of 'The New Method'"; more important for our
purposes, they predict the concerns and the metaphors of the later reading
texts. In the earlier essays "luminous details" and "donative authors" inscribe a literary history consisting of stylistic innovations and disruptions
of inherited tradition; in the 1930s the "disruptions" and "harmonies"
therein described are renamed "dissociation" and "increment of association," both of which will be considered presently. Here I would note
that despite strategic discontinuities Pound's notion of reading is crucial
to his literary criticism from the beginning; he had always viewed reading
as transformative and disseminative, not as a set of dicta to apply to any
text.

Typically, Pound's method, in *The New Age*'s Osiris essays, is neither
"new" nor "methodical"; instead, it is comprised of fragments that can
themselves be reordered by the reader, even as they describe such reorderings within literary history. Pound explains this condition of his text:
"I ask the reader to regard what follows not as dogma, but as a metaphor
which I find convenient to express certain relations" (NA X:10, 224). His
chief metaphor, "luminous detail," suggests the underlying principle of
textual production, the generation of multiple interpretations out of "interpreting detail": "In the history of the development of civilization or of
literature, we come upon such interpreting detail. A few dozen facts of
this nature give us intelligence of a period—a kind of intelligence not to
be gathered from a great array of facts" (NA X:5, 130). Not only does
Pound prefer texts which might be gathered into a palimpsest of readings,
as opposed to a continuous tradition founded on an originary poetry; he
also insists on the fragmentary and figurative qualities of his own argument. Therefore, luminous details, as though the phrase were not suggestive enough, are described by other metaphors, principally electrical circuits that " 'produce power'—that is, they gather the latent energy of
Nature and focus it on a certain resistance" (NA X:8, 178–79). The best
interpreting details, which is to say the highest art and/or the most productive readings, are nevertheless the most difficult to contain and direct;
like high-voltage wires or electronic impulses they are "swift and easy
of transmission" (NA X:5, 130).

The discontinuous history such details inscribe admits of many interpretations; history, as these early essays signify it, is a series of revo-

lutions and new inventions, not a plenum or organic whole into which innovations are simply "incorporated" or embodied (as in the Eliotic model). Pound offers a general aesthetic history—not a strictly literary one—as a collection of the luminous details that make their appearance in all his later reading texts. These details range in size and recognizability from minute discoveries of minor artists to Dante's *The Divine Comedy*. In a more discursive mode than the later essays, Pound privileges Dante, Arnaut Daniel, and Cavalcanti as interpreters who provide the sort of "marginal commentary" of Classical and canonical writers he hopes to continue in his journalistic fragments. Thus, in a transformation Eliot found particularly disturbing, Dante, the greatest of Renaissance writers, becomes (merely) "a commentator, as a friend looking with us toward the classics, and seeing perhaps into them further than we had seen" (NA X:2, 224). Pound's "symptomatic" or "donative" artists are both readers and writers disrupting and reestablishing tradition(s):

> [T]he "donative" author seems to draw down into the art something which was not in the art of his predecessors. . . . He discovers, or better, "he discriminates." We advance by discriminations, by discovering that things hitherto deemed identical or similar are dissimilar; that things hitherto deemed dissimilar, mutually foreign, antagonistic, are similar and harmonic. (NA X:8, 179)

Thus, for Pound, literary history is a palimpsest of readings, of "discriminations" and reorderings of (luminous) details. Like all culture, it is a series of individual works and authors, advancing by revising received ideas and art, thereby producing new writings which in turn generate new readings and/or writings. Pound's "New Method of Scholarship" consists of and privileges not the Great Books but discontinuous *texts* by a series of authors or authorities who are also, in the sense explored in Chapter Two, the inventors of their age. That is, they are at once creators and analysts, whose texts stand as the translative moments between cultures, not representations of a (particular) culture.

In the 1930s, reviews of Eliot's books which were written in praise of a particularly Western and Christian tradition, Pound adopts the role of "discriminator," of analyst and disturber, of (Eliot's) received ideas. To translate a phrase from the earlier essays, he is a sort of "donative reader" of the tradition Eliot was attempting to canonize. In spite of his many letters and journalistic appeals to Eliot, Pound did not succeed in influencing *The Criterion*'s editorial policy. Nor did the two old friends collaborate in a journal. Instead, they presented far different reading notions that implicitly correct each other. Eliot wrote books defining the tradition and its proper readings. Pound gathered some of his own notes into books about his reading procedure. But more telling than these are his frequent contributions to those relatively ephemeral economics and literary magazines that seem more closely to satisfy the style and editorial ideals he advanced in various communications with Eliot. It is, therefore, to a series of fragmentary reviews, replies, and "Letters to the Editor" in a little weekly tabloid that we must turn for the debate over *After Strange Gods,* a crucial debate about literary history that has virtually been lost to that history.

III

Both Eliot and, to a much greater extent, Pound contributed to *The New English Weekly,* tellingly subtitled *A Journal of Public Affairs, Literature and the Arts.* Licensed as a weekly newspaper under the editorship of A. R. Orage, *The New English Weekly* was founded to popularize Social Credit economics and only secondarily to advance an aesthetics compatible with that theory. If this newspaper comprised of interviews with prominent ministers of church and parliament, congressmen, and an occasional artist is largely forgotten, it is not because it was a unique or short-lived publication.[29] With Pound as foreign and literary correspondent (the doubly foreign correspondent writing "American Notes" from Italy to England), this newspaper lasted from 1933 to 1949, surviving Orage by fifteen years. Moreover, it was a repetition of Orage's earlier weekly, *The New Age,* an arts and economics review that published "I Gather the Limbs of Osiris" and several other early Pound essays. By strange turns of the journal trade which might long engage bibliographers and analysts of popular culture but which one must pass over in a literary essay, Orage again gave Pound control over a few pages of the weekly tabloid devoted primarily to economics, not literature. Therefore, the fragmentation of weekly serialization and the various other concessions he had to make to a nonliterary audience lend at least the consistency of repeated difficulties to Pound's elliptical critiques. Moreover, as has been suggested, Pound preferred such conditions.

At the same time that Pound was using Orage's weekly to refine notions of reading, usury, church responsibility, American history, and the function of literature in all these, Eliot was submitting articles on church polity and Social Credit to the same publication. While they do not all directly address *After Strange Gods,* Eliot's articles call for the clear definition and separation of religious and socioeconomic doctrines suggested in that book. For example, in the same issue which carried "Mr. Eliot's

Mare's Nest," Pound's first review of *After Strange Gods,* Eliot corrects editorial abuses of the proper names and territory of religion: "I should like some assurance that by 'Kingdom of Heaven upon Earth' your Theological Editor does not mean the National Dividend" (NEW IV:21, 576). Eliot voiced no substantial objections to Social Credit doctrine but wished to keep economics clearly subordinate to religion, thus to keep their languages and discourses separate. The maintenance of strict boundaries and hierarchies is a large part of Eliot's argument with Pound, and these issues can be traced in more *New English Weekly* articles than can concern us here. Yet one should recognize that the reviews of *After Strange Gods* and Eliot's responses partook of an ongoing and active argument about the proper way to read literature and create—or modernize—tradition.

Since Pound's rather disorganized reviews that comprise the debate over *After Strange Gods* are for the most part forgotten, a brief summary is in order. From 8 March to 22 November 1934, Pound wrote eight reviews of Eliot's book—though one of these is devoted to *The Use of Poetry and the Use of Criticism.* These reviews vary in length from one-half to three and one-half pages; their focus shifts from "close readings" of specific Eliotic passages to disjointed ecclesiastical histories of Medieval Catholicism through contemporary Church of England doctrine. Each review challenges Eliot "to a more precise use of language" and, on the other hand, condemns the church's "slithering away from ethical and economic responsibility," which he also identifies with imprecise terminology. Pound examines the orthodox definitions which make religion, aesthetics, and economics mutually exclusive. Exposing questionable mergers and vested interests among the three discourses in Eliot's books and in church history, Pound hopes to return to a time when "religion was real" because it acknowledged and regulated the interplay of these discourses. The titles of the reviews indicate this double focus, at once on Eliot's logic and on the whole of church history: "Mr. Eliot's Mare's Nest" (8 March), "Mr. Eliot's Quandaries" (26 April), "Mr. Eliot's Looseness" (10 May), "What Price the Muses Now" (24 May), "Ecclesiastical History" (5 July), "Mr. Eliot's Solid Merit" (12 July), "A Problem Specifically of Style" (22 November).

Eliot's replies are shorter and fewer than Pound's reviews. He hardly attempts to use them as a forum for defending himself against Pound's scattered challenges, which he generally dismisses as erroneous and irrelevant to the proper concerns of literature and religion. Of his three direct responses to Pound, "Mr. Eliot's Virginia Lectures" (15 March 1934), "The Theology of Economics" (29 March), and "The Use of Poetry" (14 June), none exceeds the half-page. Except when teased out of patience by Pound's insults, his replies are confessions of minor stylistic infelicities and citations of the same in Pound. Furthermore, several of his 1935 "Letters to the Editor" reaffirm the orthodox definitions of literature and morality from *After Strange Gods*. "Douglas in the Church Assembly" (14 February), "The Church Assembly and Social Credit" (28 February), and "The Church and Society" (21 March) show that Pound had little or no impact on Eliot's policies. One might add that Eliot succeeded in subordinating discussions of economics and politics to those of literature in his own canon and that of modernism generally. This is to say that one is surprised to find Eliot so engaged by Social Credit, even if that engagement is relegated to fragments excluded from his books of literary and cultural criticism.

For the most part, Eliot expended little time or thought on these replies. The telegraphic style and misprision of "Mr. Eliot's Virginia Lectures" are typical: "I agree with paragraphs 5, 6, 7, 8, 12, and 13, though not necessarily with every inference that might be drawn from them. . . . I fail to see their relevance to the subject of Mr. Pound's note" (NEW IV:22, 528). A point-by-point comparison of statement and counterstatement would soon lead to idle conjecture, to more "inferences" and associations than these texts can support; significantly, therewith Eliot approves Pound's most devastating attack. For example, paragraph 13 reads: "*The language of religion became imprecise,* just as the language of all forms of modern flim flam, *including popular and philological lectures.* . . ." (NEW IV:21, 500; emphasis added). Surely, as Eliot insists, he cannot sanction "every inference," but, from the Poundian perspective, only the most careless reading could have blinded him to the equation of *After Strange Gods,* a philological and theological lecture delivered in 1933 at the University of Virginia, with the decline of the church and language itself.

It was only after five of Pound's reviews that Eliot finally defended himself by redrawing the boundaries he placed around criticism—and those he tried to draw around Pound. In "The Use of Poetry" he alludes to the impossibility of "editing" Pound, a task he shared with Orage: "Now, Mr. Orage, Sir, it seems to me the time has come to engross a little more of your space to do some sweeping up after Ezra. One can't be everywhere at once; and a good deal of litter has accumulated" (NEW V:9, 213). His major complaint against Pound is that he ignores the systematic claims of *After Strange Gods* and *The Use of Poetry and the Use of Criticism*—just as journals other than *The Criterion* had privileged other systems over the moral System. Like Orage's heterogeneous little magazine, Pound's reviews fail to treat modern letters as a whole and in turn as part of the English literary "tradition," specifically what Eliot named "critical consciousness": "There was . . . one notion running through my book on 'The Use of Poetry,' something about the development of critical consciousness, which seemed to me interesting; but Mr. Pound has not mentioned it, and I do not propose to discuss it" (NEW V:9, 213). Eliot felt satisfied that his books represented the proper accommodation of "The Modern Mind" to the "critical consciousness" he traced from Dryden through Coleridge and Arnold to the London literati he published in *The Criterion* and elsewhere. One must therefore return to those more serious and well-known works for his crucial definitions and arguments, following the opposite path from the one leading through the ever more fragmentary documents of Pound's programmatic or discursive statements.

As I have been arguing all along, *The New English Weekly* does not collect the whole of the debate. Just so, Pound's first direct response to *After Strange Gods* was an eighteen-page line-by-line commentary or "close reading" of its main points. This still unpublished text, which is now housed in the Yale archives with Pound's correspondence, consists of marginal comments Pound hoped Eliot might incorporate into his revision when the lectures were published in an English edition (they were first published in the United States under contract with the University of Virginia).[30] This seems to be strategic on Pound's part, for he sent Eliot this long review with an apology for Orage's instigations: "I have toned down my think [sic] about yr/Heresy (alleged Primer which it AINT) cut from 18 pages to dignified remarks on 3 idem. . . . Orage evidently wants an argument." In the same letter, he offers to correct Eliot's logic and challenges him to reply in *The New English Weekly:* "Seems to me it wd be better for you to reply to my general statement with something more than the lectures which contain non sequiturs."[31]

Ironically, Pound's reviews make non sequiturs where none (would seem to) exist in Eliot's argument. He usually plays on the fringes of Eliot's book, exposing (in)significant minutiae and grounding his attacks on unauthorized suggestions and errors.[32] Sometimes this strategy uncovers Eliot's concealed purpose and commitments; sometimes it simply enables Pound to turn the argument to his own preferred topics. He manages both when he opens his first review with a (re-)statement of Eliot's "thesis" discovered on the dust jacket. In this way, he makes the whole argument rest on words probably written by an editor and remaining with few of the extant copies of the book:

> Mr. Eliot's thesis, nowhere so clearly expressed in his text as in the jacket announcement, is that: "The weakness of modern literature, indicative of

the weaknesses of the modern world in general, is a religious weakness,
and that all our social problems, including those of literature and criticism
begin and end in a religious problem." (NEW IV:21, 500)

That attributed thesis statement is closer to the concern Eliot expressed
in his editorial statements in *The Criterion*, where he was interested in
defining the sociology of literary criticism as such. In *After Strange Gods:
A Primer of Modern Heresy*—a title that stuck despite Pound's quibbling
with its pedagogical ("primer") and innovative ("modern") intentions—
is aimed at the unorthodox critical practices of individual artists.

Moreover, while Pound's reviews do address the privilege Eliot afforded religion (or religiousness) to the virtual exclusion of economics
and politics, this topic is subordinated to that of practicing (though not
proposing, as one would a topic) his reading against Eliot's tradition.
Therefore, as though to repeat the lessons of "How to Read," he "corrects"—or, better, distorts and supplements—Eliot's book with his own
readings ("diagnoses" of modern culture) of Eliot's corrective readings,
and his notions of "dissociation" and "increment of association." Recall
that, whether or not Eliot had "How to Read" in mind, *After Strange
Gods* proscribes the very sort of reading that text represents. Thus, notwithstanding a certain indirection, Pound reacts to the following charges
which comprise Eliot's thesis statement:

> *I am not here occupied with the standards,* ideals and rules which *the artist
> or writer should set before himself,* but with the way in which his work
> should be taken by the reader; *not with the aberrations of writers, but with
> those of readers.* (ASG, 24; emphasis added)

Over the course of his reviews, Pound distinguishes his "aberrations"
from Eliot's definitions of the proper functions of "readers" and "writers." Characteristically arguing in a manner that confuses these two functions, he *reads* (as one might give a close reading of a poem) Eliot's
book in such a way as also to present his own abusive reading procedure.
All this Pound collects into his aphoristic descriptions of "dissociation"
and "increment of association" placed at strategic points in the reviews.
However, since he puts these notions into play (one might say "into use"
or an acceptable "usury," as the French *usance*) by naming them, one
cannot simply place citations of these against Eliot's more straightforward
definitions. This sort of interfiling or interfacing which Pound sometimes
effects against Eliot's book makes little sense out of context—a context
to which Pound clearly intended to wed his commentary. Furthermore,
one might turn Pound's strategy into a doctrine—a trope he successfully
resists. Such a methodological investment not only ignores the fragmentary style (not merely the "fragmented condition") of Pound's text; it

reinscribes Eliot's orthodox readings and *orderly* argumentation. Therefore, one should defer assigning fixed meaning to Pound's scattered counterthesis until following through its workings against Eliot's text.

Out of Eliot's virtual refusal to defend himself, Pound makes a pretext for the sort of lengthy explanations he generaly abjured in his criticism. For example, when Eliot innocently asks him either to avoid or define such vague judgments as "when religion was *real*," Pound answers with the following revisionary reading of Medieval art and religion:

> *Let me rewrite it thus:* During those centuries when organized Christianity, namely, the Roman Catholic Church, was most active in the life of Europe, both in affairs spiritual and affairs temporal, *during those Ages when religious verbal manifestations in Europe reached their most admirable heights,* whether in the writing of Scotus Erigena, Albertus de la Magna, Aquinas, Francis of Assisi, Dante himself; *and when ecclesiastical architecture triumphed* in San Zeno, St. Hilaire, the Duomo of Modena and an infinite number of churches, *the CHURCH had not abrogated her right to dissociate ECONOMIC right from Economic evil.* (NEW IV:24, 559; emphasis added)

Departing thus from a rewriting of his own imprecise phraseology, Pound rereads, which is to say rewrites, aesthetic and religious history. Nearly every time he turns to revise either his own or Eliot's text, he (con-)fuses the activities of writers and readers Eliot tried so carefully to distinguish. Such a substitution is really an economic exchange, not a simple reversal but a gain or loss of meaning and force. Coupled with his questionable choice of Eliot's thesis and main points, a revisionary strategy of assigning different meanings, histories, and values to Eliot's words allows Pound substantially to rewrite *After Strange Gods*. Taking advantage of the reviewer's disadvantage in not having the text in common with his own readers, he quotes only the smallest fragments from Eliot's book, especially those words which resist final definition. In place of summaries of Eliot's argument, he offers heretical readings of crucial terms employed by Eliot in the most orthodox (critical) sense. Further, throughout the reviews, by way of *defining* (one will come to recognize this procedure as "dissociation") his and Eliot's crucial terms, Pound shifts from specific details to nearly unreadable generalizations. As though to present additional words that must undergo the same process of interpretation he applied to "when religion was real," he ends with assertions that are either capable of infinite revision or simply too vague to have any meaning at all. Perhaps no phrases are either as full or as empty of meaning—depending on how far one is willing to go in gathering the relevant Poundian and Eliotic associations—as "ECONOMIC *right*" and "Economic *evil*." This marks only one of Pound's refusals to answer requests for clear definitions. Instead, he forces Eliot and us further into the interpretive enterprise he practices.

The search for definitions and meaningful contexts for such fundamental assumptions as "right" and "evil" (re-)activates Pound's reading procedure, setting in motion an "increment of association" which can become an uncontainable dissemination of meanings and examples—especially as it acts upon and reacts to Pound's more well known critical dicta. These associations again disturb distinctions between reader and writer, art and the other discourses. And Pound's examples affirm, rather than halt, these transformations and translations. One can again use the above citation. Note that "Dante himself" is one of a list of theologians, not poets, who presumably represent "real religion" or the Catholic Church as regulator of language "both in affairs spiritual and affairs temporal." Perhaps no other poet is more open to interpretation than Dante—notwithstanding Aquinas's formalized hermeneutics. Pound's reading of Dante, not as the finest poet of an age when theology and aesthetics were one (Eliot's reading), but as economist and reader, *instigates,* in Pound's very special sense of that verb, the very sort of adulteration Eliot banned from his critical orthodoxy. For instance, he objects to *The Cantos'* reading of Dante's *Inferno:* "It consists (I may have overlooked one or two species) of politicians, profiteers, financiers, newspaper proprietors and their hired men. . . . It is, in its way, an admirable Hell, 'without dignity, without tragedy'" (ASG, 46). Eliot's reading of *The Cantos,* not merely his request for clarification of the review's vagaries, motivates Pound's revisions.

Ignoring the certain potential for anarchy in his reviews, Pound abbreviates the list of his privileged writers from his "true hierarchy of values" elaborated in "How to Read" and elsewhere. He honors those artists who produce the highest "verbal manifestations," a category that grows to include musical, scientific, artistic, theological, and, as in our passage, architectural forms or signs. All these consist of words and other structures which found and characterize—even read—the Great Ages of (art) history. In Pound's strange (inflationary) economy of multiplying adjectives and activities, the list of associated discourses grows endlessly. But as the adjectives increase and change, the proper nouns, Dante and Erigena and Cavalcanti and the others recur. In 1933, Eliot complained of Pound's admiring Cavalcanti for the heretical pratice of "some pneumatic philosophy and theory of corpuscular action which I am unable to understand" (ASG, 45). One also recognizes Pound's heresies in his earliest revisionary histories, including "I Gather the Limbs of Osiris," that list representative artists who go beyond their own art and inherited artistic traditions to disrupt and finally to define their age. Thus, not only does Pound confess "heretical" individualism and violations of the proper boundaries separating literature from the lesser genres; he finds precedents for his procedure within the most secure parts of (Eliot's) tradition. The greatest poets of the past become representative revolutionaries, if you will, seemingly endorsing the very things Eliot decried in modernism: "extreme individualism in views, and no accepted rules or opinions as to the limitations of the literary job" (ASG, 34).

In the middle of this series of reviews of *After Strange Gods,* Pound writes an uncharacteristically direct review of *The Use of Poetry and the Use of Criticism,* in which he expands his list of privileged readers to include several extraliterary critics particularly unacceptable to Eliot and the literary establishment. This review, "What Price the Muses Now," elicited Eliot's longest and most serious reply in the form of a defense of "the modern mind" against the chaos of facts and opinion that constitute Pound's reading procedure as Eliot viewed it. Pound opens this exchange by trying to place—or win—Eliot into the circle of "revolutionary" readers, those minor figures from whom he amalgamated the "method" Eliot found heretical. He names Eliot a subtle revolutionary for his then canonical readings of seventeenth-century poetry: "Possibly does not seem revolutionary to inexpert auditor or the layman, but supposing an acquaintance with academic or professional opinion of university specialists, the essay is full of startling divergences from the accepted opinions of this body of men" (NEW V:6, 131). Supposing that Eliot's readings of individual poems are revolutionary, and no one can deny their occasional brilliant originality, his aim was the conversion, not the overthrow, of academic criticism. After the unsuccessful rhetorical maneuver of flattering the opponent, Pound tries a string of abusive *ad hominem* attacks upon the modern literary critics Eliot canonized as representative of "the modern mind": "The gravest charge against his rank as critic is that he ventures to label an essay 'The Modern Mind' [the final chapter of *The Use of Poetry*] and to have found NONE of the more vigorous intellects of the past century worth even passing mention" (NEW V:6, 131). In the attack on Eliot's critical models, Pound finds Jacques Maritain "the typical French religious faddist, and a racketeer on the borders of aesthetics" (NEW V:6, 131). Not only does Pound fail to define his differences with Maritain; he seems merely to return to Eliot the charges of blurring the borders of separate discourses. He attacks Maritain's sanctioned fusion of religious and literary judgments with invective that comes from the lexicon of the very orthodoxy he would seem to criticize. Maritain is a "faddist"; Eliot's praise of him is a "nasty blasphemy." Pound's own use of language would seem uncritical, even as it succeeds in marking similarly uncritical moments in Eliot's text.

In a more positive vein, Pound lists his own "heterogenius" (a pun/typo/misspelling lost between the typescript and published versions of "Mr. Eliot's Solid Merit") precursors and allies from whom he sometimes claims to have borrowed *Anschauung,* if not method: "His [Eliot's] highwater mark is an allusion to Levy Bruhl, who is at any rate a first-rate professor. *Fabre, Fraser* [sic], *Frobenius, Fenollosa would appear to mean nothing to him*" (NEW V:6, 131; emphasis added). Suffice it to say that Pound's list of modern readers hardly absolves him of charges of individualism and asystematism. Instead, it is positive proof that he valued detail over system and, worse, metaphor over method. From Fenollosa he had the "ideogrammic method," which is among other things a metaphor for poetry and a call for a more metaphoric poetics; from Frobenius he borrowed the metaphors of *Kulturmorphologie* (which he took to be a cultural tropology), *Paideuma,* and "increment of association." Fabre, like the anthropologist Frazer, catalogued fragments, and his entomological catalogues present analogues to the cultural principles Pound read in more obviously aesthetic works. Like all his model readers, they offered new directions in the concern of lost histories and associations of words Pound thought were stored, like centers of energy, in (poetic) texts.

Perhaps he expects Eliot and other readers to be familiar enough with his own works and with those of his model readers to make sense of his position from that list of names. In any case, he does not summarize either his position or those of his sources. This indirection provokes Eliot to clearly define, at least by exclusion, the "modern mind." Ever the editor, Eliot attempts to correct Pound's spelling, but his principal aim is to correct Pound's preference for heterogeneous details over a unified tradition:

> Mr. Pound suggests that "Fabre, *Fraser,* Frobenius, *Fenellosa*" [emphasis added] mean nothing to me. *I at least know how to spell Frazer,* [emphasis added] but beyond that, Fabre and Frazer mean to me very valuable collectors of facts. Fabre was, so far as I am qualified to judge, a great observer, Frazer a great collator; but neither of them was, so far as I know, and in the sense which I was using the word *mind,* a great exemplar of the modern *mind.* (NEW V:9, 213)

Finally addressing the details of the reviews, Eliot reduces Pound's argument to spelling errors. He had neither the patience nor the doctrinal tolerance to grant Pound a countersystem, let alone an epistemology that abolishes system and Mind. In *Guide to Kulchur,* Pound says, "All knowledge is built up from a rain of factual atoms" (GK, 98). While the radical challenge posed by that aphorism cannot be explored here, one should note that Eliot's own spelling error marks the eccentricity of Pound's contrary modern canon. This is perhaps to belabor what might be a simple error, a slip of the typewriter or the typesetter, but Eliot misspells Fenollosa. This would be insignificant if it weren't for the fact that Eliot, by virtue of *The Criterion* and Faber & Faber, had the power to determine which names would be common enough to be spelled correctly—at least by those presuming to write literary criticism.

If Pound failed to note this particular accident in opening his next review with the joke "Possum tree'd with a spellin' book," he delighted in exposing Eliot's habitual reliance on orthodox arguments and other basic tools of the serious professor. Just as "Mr. QT slumbering over his grammar book" is offended by Pound's lacunary style, Possum would quiet his old friend with a spelling book. If he carefully avoids such books himself, Pound's readings of Eliot often *analyze* (or, better, abuse) a word until it comes to bear more meanings and associations than could have been intended and can be controlled. His most frequent strategy is the exposure of assumptions grounding the key words of Eliot's argument: "heresy," "revolution," "orthodoxy," "modern," and "religion." In vain, he would "lead Mr. Eliot to a more precise use of language" (NEW V:4, 96). Focus on the specific details and words of Eliot's argument makes Eliot appear guilty of the very slippages of language Pound at once deplores and uses. As he reads *After Strange Gods* and the general context of those lectures, Pound does not offer more precise terminology; instead, he disturbs orthodoxy by inhabiting its very language, its discrete words and phonemes.

In this way, Pound announces his own *revolution*—at once a verbal troping and an institutional overturning—in a pun: "The difficulties of a suppressed *author* in dealing with an accepted living *authority* are almost insuperable" (NEW IV:24, 558; emphasis added). From the perspective of more rigorous postmodern tamperings with the notion of authorship, Pound's pun appears the sort of "worn-down" figure he condemned as the "usury" of overused rhetoric, but such tropes concisely expose his distance from Eliot and the latter's new role as general authority. Still, Eliot felt compelled to say: "I am uncertain of my ability to criticize my contemporaries as artists; *I ascended the platform of these lectures only in the role of moralist*" (ASG, 10; emphasis added). That appears in the same book that criticizes the modern tendency to break the "accepted rules" about the "limitations of the literary job." There is, of course, no contradiction, for Eliot is affirming, not violating, orthodoxy. The reference to the decidedly Anglo-Catholic, which is to say the moral, God can be taken nearly for granted—if one can proscribe modernism's "Search for Strange Gods," Eliot's titulary promise. Pound will not grant such assumptions; this refusal is one of the things that earned him the special status of "heretical reader."

In marked contrast to Pound, who had always to begin again each time he wished to present or explain his heterogeneous critical notions, Eliot frequently appeals to beliefs he shares with his audience, including his own critical terms and formulations. Pound's readings depend on exposing the centrality and the uninterrogated (in-)stability of such assumptions. Therefore, and necessarily by indirection, he asks whether Eliot's placing of the questions of literature within religion or morality—his aestheticizing of Anglo-Catholicism, or is it Anglophilia?—truly addresses the question of culture and its decline especially when it excludes those secular discourses and "contemporary artists" outside the scene of Eliot's concern, the nexus of the university and church.

Pound's remarking of Eliot's slide from author to authority is not an isolated attempt to call Eliot back to a more radical modernism. His goal is not so much to correct Eliot's expansion of the limited role of author as to protest the limits he placed on literary criticism—the orthodoxy and tradition he imposed on reading. Pound is most effective when he can turn Eliot's words against themselves in order to expose an iconoclastic— even an American—Eliot. Sometimes he uses jokes, puns, malapropisms, and other devalued figures to accomplish his own serious aims. For example, by recalling his old-friend's earlier impieties, he tries to undo the exlusivity and elitism of modern English—which is not to say Amurikan—letters: "Mr. E[liot] once said to an etiolated epigon [sic] of the British Upper Middle reaches 'What you need is a plate of good buckwheat cakes.' It is perhaps time to make similar abrupt suggestion to the *dean of English essayists:* What you need is a bit of good solid reading matter" (NEW V:6, 132; emphasis added).

Pound also reminds Eliot of his own pedagogical responsibility to inform his American audiences, the Harvard and Virginia students and faculties as well as the wider reading public, of the reprehensible state of literature and criticism in their country. Picking up a favorite theme of translating revolutionary European art (back) into America—of grafting an aesthetic onto a political revolution via a transatlantic economic-political-cultural exchange—Pound outlines the various targets of his own reading procedure. Allegedly for the benefit of the few "intelligent students who erred into Harvard," he supplements the introduction to *The*

Use of Poetry and the Use of Criticism, first delivered as the Charles Eliot Norton Lectures of 1932–33:

> His opening pages would . . . *if properly used, serve to clear off a good many vermin,* partly British and mostly long-standing, whereas the mentally alert sufferers at Cambridge, Mass., had hoped that the more immediate obstacles, cankers, barnacles on *American literary life would be scraped.* They felt let down by the lecturer's dealing with faint and far away whiffs of Dryden and Mat Arnold. . . . These youngsters having the weekly and monthly spew of Canby and Co., the "Atlantic," a quarterly affair disrespectfully referred to as "Bitch and Bugle" ["The Hound and Horn," Blackmur's Harvard review], and the old line demo-liberal tosh of the N.Y. weeklies, copied and dishwatered down from the very stale stink of the London weeklies, etc., had hoped that *the dean of English Criticism and Editor of Britain's Brightest Quarterly would fire a few volleys at some of the more overshadowing pests.* (NEW V:6, 130; emphasis added)

In this way, Pound hopes to destroy (his words are "clear" and "scrape") various kinds of literary critical orthodoxies. Elsewhere his critical hygiene uses the metaphors of pouring acid and performing surgery, but one might add that his destruction is also an inhabiting and a subversion, a de-*con*-struction. Even when it lacks the rigor of self-recognition, it manages to de-structure the very canon which Eliot's book was written to establish. It is fantastic indeed to imagine that "the dean of English Criticism" might be engaged to prevent the influence of English criticism and literature on America. But, despite a certain inconsistency in his own argument, Pound there lists the grievances which comprise the general context of his reviews.[33] It is noteworthy that Pound again—and concisely—refers to his own wider project of establishing a definitely American canon, even an American criticism. And, more to the point, he does so by positing a "mentally alert" audience reminiscent of that rare "sane and intelligent reader" who might prefer Pound's lacunary style to Eliot's. Thus Pound abuses the intentions, text, and audience of the "dean of English criticism," effecting the sporadic attack upon the total reading scene of Eliot's criticism, if not totalizing a reading of Pound's own countertradition.

If the great Dryden-Coleridge-Arnold-Eliot tradition of poet-critics remains intact, Pound nevertheless suggests a way in which Eliot might be "properly used" against himself. That proper use is clearly an abuse, one decidedly different from the borrowing and wearing down that Pound condemns as "usury." If the pun on use—on the proper use of, or borrowing from, tradition—was not strategically deployed in the above passage, Pound's reading procedure, which he later identifies as a "productive exchange" of meanings, adds unintended, even absurd, meanings to Eliot's text that seem to have been there all along.

Pound insists that Eliot is intentionally vague and imprecise in his choice of words, that he employs "the mathematical fallacy or trick equation of using zero or infinity as the middle link in a proof" (NEW IV:24, 559). In the interests of ecumenicalism, of appealing to an audience, Eliot had merely cited "religion," not a specific religion, as cure for modern moral decay. Pound would seem to address a whole series of assumptions that allow Eliot to rest his argument on abstractions and generalizations. After all, one can simply take as a given the moral code that underlies the Western Judeo-Christian tradition. Eliot's own subtle redefinition of "tradition" depends on this unquestioned ethnocentrism: "What I mean by tradition involves all those *habitual* actions, habits and customs, from the most significant religious rite to our *conventional* way of greeting a stranger" (ASG, 18). Pound accuses Eliot of using a general notion of religion in order to avoid the economic and political problems that he should address: "Mr. Eliot's book is pernicious in that it distracts the reader from a vital problem (economic justice); it implies that we need more religion, but does not specify the nature of that religion. . . . I mean Mr. Eliot does not discriminate in favour of a kind of religion that might be beneficial to men's minds, manners or morals, or to social amelioration" (NEW IV:24, 559). Drawing Eliot's argument away from his Anglican and anglophile concerns into the wider area of religion but continuing to cite specific words and unaccountable details of his presentation, Pound at once prepares the way for his own revisionary church history and exposes the underpinnings of Eliot's authority. He fixes on questions which Eliot and fellow members of his poetic and theological orthodoxy ignore in the interests of maintaining tradition and advancing (properly philosophical) arguments. After all, how many readers stop to ask "Which religious denomination?" "What literary tradition?" "Why Great Books?" The traditional canon and curriculum are things one merely departs from.

Indeed, Eliot is quite frank about refusing to engage the radical questions Pound happens upon with regard to the unstable categories of literary and religious orthodoxy. Pound's mention of the "trick equation of using zero or infinity" alerts one to Eliot's reservations, his authoritative avoidance of controversy and discussion:

> I do not wish to preach only to the converted, but. . . . *In our time, controversy seems to me, on really fundamental matters, to be futile. It can only usefully be practiced where there is common understanding. It*

requires common assumptions and perhaps the assumptions that are only
felt are more important than those that can be formulated. (ASG, 11; emphasis added)

One might consider that an explanation of why Eliot neglected thought-
fully to respond to Pound's reviews: since Pound refused to grant such
common assumptions, there was little hope of agreement, let alone of
cooperation in mending the commonplace absence of a common tradition
that is the special inheritance of modernism.

Because it cannot be brought fully under conscious control, the province of tradition must be carefully policed. "Orthodoxy," both as a general principle of acceptability and as specific dicta, provides enforcement.
This militant conservatism is the most significant addition *After Strange
Gods* makes to the formulation of tradition in "Tradition and the Individual Talent":

> [A] *tradition* [emphasis in text] is rather a way of feeling and acting which
> characterizes a group throughout generations; and that it must largely be,
> or that many elements in it must be unconscious; whereas the maintenance
> of *orthodoxy* [emphasis in text] *is a matter which calls for the exercise of
> all our conscious intelligence* [emphasis added]. (ASG, 31)

Orthodoxy is maintained, not by reflection on its own language and procedures, but by strengthening institutions which preserve and communicate traditional values: universities, the church, the literary canon and
its interpretations. Faced with innovations that must either be excluded
or accommodated, orthodoxy prescribes standards for determining what
will be subsumed into the privileged category "tradition." As Eliot says:
"Tradition has not the means to criticise itself; it may perpetuate much
that is trivial or of transient significance as well as what is vital and permanent" (ASG, 31). Just as there is something circular, or at least self-
fulfilling, in the judgment that certain texts (little magazines, letters,
aphorisms, fragments) are ephemeral, orthodoxy not only determines value,
it can doom unmanageable details to transience—it can, that is, if it can
banish heretical readers who continually unearth and prize such heterogeneity. And what could be more transient or trivial than the details of
literary or ecclesiastical history on which Pound grounds his readings?
What more insignificant than Dante's economics or Eliot's revolutionary
tendencies?

However, it is the devalued, the insignificant details of his own argument (the "common understanding" amongst the orthodox which makes
"controversy futile") that expose Eliot to Pound's reading procedure. It
is Pound's tamperings with details of language, as much as the larger

interpretive issues of modern readings of Dante, that the Eliotic argument cannot accommodate. Moreover, the focus on individual letters, prefixes, and unusual definitions strategically opens Eliot's book to questions of reading and economics—of the reading economy—that are Pound's principal concerns. Departing from the etymologies and affiliations of individual words, he analyzes Eliot's grounding concepts: "tradition," "orthodoxy," "heresy." These, he discovers, are mistranslations from the debased theology that had abrogated ethical and linguistic responsibility.

Refusal to argue in Pound's terms implicates Eliot in the alleged decline of the Catholic church from its Late Medieval acme, when argumentation and precise terminology were valued. Against literary and theological orthodoxy, Pound offers the following revisionary history of the fall away from heterodoxy:

> Time when church no longer had faith ENOUGH to believe that with proper instruction and argument the unbeliever or heretic could be made to see the daylight. *Invocation of authority to MAKE him believe.* This runs up to debate (correspondence) between Leibnitz and Boussuet (*Leibnitz can be spelled without the t if Mr. Eliot desire*). (NEW V:12, 272; emphasis added)

That parodies Eliot's authority, his invocation of known authorities, and his imposition of arbitrary rules such as the spellings of proper names. More importantly, Pound concisely suggests the necessity of attending to language.

Significantly, the Catholic church's neglect of language and debate which Pound underscores coincides with the rise of philosophical abstractions, another linguistic disease addressed in the critiques of Eliot and crucial to Pound's poetics at least as early as Imagism. Yet one should recognize that Pound does not simply prefer a "natural language" to abstractions or sophistry. He does not damn rhetoric; instead, his call for more careful argument mandates a language at once more suggestive or metaphoric and persuasive, a language capable of many interpretations and of provoking interpretation yet not indefinite. Rather than bringing Eliot's language "closer to the thing" or into agreement with proper usage, he shows that Eliot has abused certain key words in order to produce (the illusion of) univocity where it does not exist, an orthodoxy where there is only heterodoxy and heterogeneity. It is in the dialogue of the latter that Pound finds the "live tradition" and "increment of association" that will link modernism with the energies of cultural origins. His major strategy is to lift Eliot's words from their context as part of the invisible "authority" of received ideas attributable to canonical authors. He cites imprecisions, but more to affirm the freedom of language than to suggest ways Eliot might make his words and hierarchy more rigid.

Now we can turn to the minute workings of Pound's reading procedure, the *analysis* (which one will come to recognize as "dissociation") of individual words. Pound accuses Eliot of equivocating on "orthodoxy," ironically the very name of univocity. Eliot clearly uses the word in the general, approbatory sense of agreement with standards; any adjective might precede this use of orthodoxy: "religious," "literary," "political." But, perhaps in keeping with his own notion of the moral, and decidedly Christian, role of literature, Eliot also uses "orthodoxy" in the sense of the only permissible theological doctrine. He has a good deal of difficulty with his terminology at this point, seeming almost to invite Pound's objections:

> With the terms ["orthodoxy" and "tradition"] in their theological use I shall presume no acquaintance; and I appeal only to your common sense; or, if that sounds too common, to your wisdom and experience of life. That an *acceptance of the validity of the two terms as I use them should lead one to dogmatic theology, I naturally believe; but I am not here concerned with pursuing investigation in that path.* (ASG, 33–34; emphasis added)

Eliot's borrowings from the theological lexicon affiliate him with dogmatic theology anyway. Pound, who presumed more than a passing acquaintance with theological jargon and etymologies, forces him down the very dogmatic path he then decries.

Pound's most successful assault on Eliot's terminology occurs in his ironic, oblique, but double-edged defenses against charges of heretical interpretations. Recall that Pound's own heresy involves an alleged preference for Cavalcanti over Dante, that "Guido was very likely a heretic, if not a sceptic" (ASG, 45). His reviews in *The New English Weekly* repeat his earlier positioning of Dante as heretic in whose texts there are crossings of poetry and politics, religion and economics, prose and passion. Furthermore, the unpublished version explores Eliot's strategic misusages of "orthodoxy" and "heresy":

> I do not see why he need "suspect me of finding Cavalcanti more sympathetic than Dante." I thought I had made my preferences quite clear. And after using heretic and orthodox in very general senses, he jumps to the perch of *heretic as one who couldn't have a certificate of orthodoxy from the college of cardinals at a particular date.* (Unpublished review, p. 13; emphasis added)

The difference between—which is also an affiliation of—heresy as a dissenting opinion or practice and as a punishable offense is crucial to Eliot's argument and has profound implications for the modern canon. After all, he is, as it were, passing out certificates of canonization (in the literary if not in the sacerdotal sense) to modern writers who might or might not be published or become part of the university curriculum. Translating a historical paradigm of literature or art into theological terms, he equivocates a second time and again causes his argument to rest on a shaky word, even two: "*orthodoxy* . . . seems to me more fundamental (with its opposite, *heterodoxy,* for which I shall also use the term *heresy*) than the pair *classicism-romanticism* which is frequently used" (ASG, 22). If Pound failed to cite that example, he pointed the way back to Eliot's specific words and behind his vested interests.

After Strange Gods bases its own readings on an equivocal use of the pejorative phrase "heretical readers." Eliot sometimes means heresy in the (almost) purely literary critical sense of, say, "the heresy of paraphrase." He also ranks D. H. Lawrence, Katherine Mansfield, Pound, W. B. Yeats, and James Joyce in ascending order from the worst heretic to the most (Roman) Catholic, this according to their treatment of moral struggle and their depictions of the Christian Hell. Because it provokes Pound's most flagrant abuse ("misreading") of Eliot's orthodoxy, one might recall Eliot's charge that "Lawrence is for my purposes an almost perfect example of the heretic" (ASG, 41), because a widow in "Shadow in the Rose Garden," struggling with good and evil fails to imagine a Christian afterlife. Eliot objects to Lawrence's psychologisms (Freudian jargon) and to his characters on the grounds that they lack "any [Christian] moral or social sense": "the characters themselves, who are supposed to be recognizably human beings, betray no respect for, or even awareness of, moral obligations, and seem to be unfurnished with even the most commonplace kind of conscience" (ASG, 39–40). Eliot attributes Lawrence's heresy to a failure to keep literature properly within aesthetic and theological boundaries.

Commenting on that passage (only the unpublished version cites it as such), Pound reads (Eliot's?) orthodoxy, not as a limiting principle, but in such a way as to stretch its meaning to cover his own cross-discursive practice. He ends the first review of *After Strange Gods,* "Mr. Eliot's Mare's Nest," with what might appear, in the absence of textual grounding in Eliot's ambiguous argument, a shocking misreading of the very strictures Eliot placed against literature's encroachment on psychology, politics, and economics. The reading of "orthodoxy" is nearly what Eliot meant by "heresy":

> It is highly confusing to find half way through the book that what he means by "orthodoxy" is merely the extension of literary subject matter to certain

ranges of human consciousness that inferior writers now neglect as have inferior writers in other times. (NEW IV:21, 500)

Thus Pound defines—in fact dissociates—his own sense of literary orthodoxy, which (abusing Eliot's intentions) includes his own heterogeneous economic and political concerns: "In Dante's intellectual world certain financial activities are against nature" (NEW IV:21, 500). He does not simply abandon the notion or the word "orthodoxy." Instead, he takes it apart and then forces it to bear his own meaning. This is consonant with his dissociative reading procedure, which can be seen to share a good deal with Nietzsche's transvaluation of Wagner's philosophemes.

Over the course of the reviews, Pound erects dissociation into an economic and interpretive principle in opposition to Eliot's manner of defining words and literature. An argument over the proper use and relative virtues of clear definition and dissociation underlies their repeated calls to precise language throughout the *New English Weekly* exchange. As has been suggested, Eliot wants Pound to stick to the point and avoid the vague polemics of words like "good" and "bad," while Pound fragments the apparently stable definitions of Eliot's key terms. Indeed, though they do not surface as such in these reviews, Pound's and Eliot's conflicting definitions of "definition" underwrite these debates. Therefore, before treating the word and procedure of dissociation, which is one name for Pound's rather unwieldy manner of definition, one should recall Eliot's own strategic definitions, which are part of the traditional critical aims of assigning stable meanings to words and fixed boundaries to the categories within culture. Nowhere is Eliot's vigilance in rectifying heresies against literature and philosophy greater than in *Notes Toward the Definition of Culture,* where even his epigraph, tellingly drawn from that modern institution, the OED, rewrites his title and abbreviates the thesis he argues here and in *After Strange Gods:* "DEFINITION: 1. the setting of bounds; limitation (rare)—1483—*Oxford English Dictionary*" (DC [1]). Such is Eliot's (and the onto-theo-*logical*) expectation for individual words and interpretations of literary or philosophical works within the aesthetic tradition. Clearly, tradition and orthodoxy are this sort of limiting definition, despite their failure fully to withstand Poundian analyses.

Pound's writings and readings always strained against the limitations or boundaries Eliot imposed on (modern) poetry and the lesser genres of art and criticism. While he never addresses Eliot's manner of definition as such, he advances various specific dissociations in opposition to it. Pound borrowed the rhetorical (even Nietzschean) strategy of dissociation from Remy de Gourmont, whose use of that metaphor is far from straightforward.[34] In fact, one is more likely to associate it with Eliot's pejorative "dissociation of sensibility," also a reading of Gourmont. Eliot's more familiar formulation in "The Metaphysical Poets" names the seventeenth-century breach in the classical unity of thought and feeling, poetry and life. Eliot laments these dissociations as the irreversible if not Biblical fall into the Romantic opposition of "intellectual poet" and "reflective poet" (SE, 247) which initiated a continuing decline into the very sort of heterogeneity and fragmentation (modernism) Pound seems to represent, if not to privilege.

For an understanding of Pound's far different interpretation and employment of dissociation, one has recourse to Gourmont's celebration of the transformative—even translative—effect of all language, in "Dissociations" (*Le Culture des idées,* 1900), an essay Pound frequently employed though, characteristically, failed to cite. In this text, Gourmont, rather than simply defining the metaphor of dissociation, displays its workings in sexual metaphors; these suggest what Pound elsewhere called the disseminative force of reading—of all language:

> It [dissociation] is a matter either of conceiving new relationships among old ideas and images; or of separating old ideas, old images united by tradition, and reconsidering them one by one, being free to rework them and arrange them in an infinite number of new couplings—which a new operation will disunite once again until the new ties, always fragile and equivocal, are formed.[35]

For Pound, as for Gourmont, dissociation is a strategy as well as the natural condition of language, not some calamity that befell an original unity or the cultural plenum. If anything, language has worn down from (Pound's phrase is "slithered away from") its analytic or interpretive function into abstraction and generalization; hence Pound's condemnation of philological argument as "bad rhetoric" or the "vague middle" (and sometimes "muddle") of the clichés and dead metaphors of some modern poetry and criticism. Again, this argument is as old as Imagism.

In the reviews of Eliot as well as in other readings of the 1930s, Pound uses dissociation in at least three ways: as a procedure for defining (or, better, de-defining) individual words and positions, as a shorthand manner of introducing data for analysis, and as the name for, the principal tool of, the sort of "productive exchange" he sought in reading and the larger economy of textual exchanges.

In the first instance, covered by Gourmont's metaphorical definition, dissociation works through the surface of language to fragment, agitate, and otherwise *read* established relationships between ideas and things, words and grammar. Carefully employed, dissociation yields minute analyses of commonplace notions, clichés of thought as well as clichés of language. This sense of dissociation we have already seen working against Eliot's "orthodoxy" and "heresy."

Dissociation can also detail the central words or texts to be treated in an essay or other argument. The unexplained quotations and aphoristic snatches of method ubiquitous in Pound's reading texts and in the "editorial program" of *The Exile* exemplify this sort of dissociation which might also be termed "disjunction" or "juxtaposition." Pound makes this usage explicit in his introduction of his favorite economics quotations and definitions in *The ABC of Economics:* "1. *DISSOCIATIONS: Or preliminary clearance of the ground*" (SP, 233).

Somewhat apart from his rigorous yet elliptical usages, Pound also employs "dissociate" in the colloquial sense of "to distinguish" unequals or "to disengage" false identities. In "Mr. Eliot's Quandaries," the second and longest review, he offers the following dissociation, bearing on both Eliot's style of criticism and Pound's overall reading procedure:

> I would like to *dissociate two sets of writers;* the first, *those who, when possible, try to focus debate on the "truth itself,"* or (to avoid great phrases) upon the right answer to any question under discussion, *rather than on the errors committed by Mr. So and So.* (NEW IV:24, 558; emphasis added)[36]

This is to say that, for the most part, Pound focuses on reading Dante, et al., in the most suggestive ways, rather than on the *heresies* of others. If one takes that polemical dissociation at face value, as consistent with (a New Critical, even a neo-Aristotelian) Pound who devalues criticism in favor of the concision and directness of poetry, then the profusion of commentary and perspectives he writes in the margins of *After Strange Gods* is simply erroneous. Even if Pound held out hope for a "truth in itself," it was a truth capable of many interpretations. Just so, he prefers to exploit historical periods and texts that are dissociative (or "interpretive" as "luminous details"). Even after his dissociation from Eliot, his dissociative readings intervene to disturb the stable meaning of his own key words: dissociation dissociates.

Moreover, all history, but especially ecclesiastical and literary history, is a decline from dissociative language and arts into abstractions erected to disguise or halt the freedom and play of language and art. Again, one has recourse to Gourmont's more discursive formulation of Pound's practice. For example, in a passage reminiscent of the Poundianism that claims one can judge the degree of *usury* in a civilization by the distinctness of line in painting and architecture ("with usura is no clear demarcation" [XLV:229]), Gourmont suggests that the greatest periods of history are *dissociative,* that they used language interpretively: "One could undertake a psychological history of the human race by determining the exact degree of dissociation which characterized, in the course of centuries, the number of truths the right-thinking people agreed to call primordial (SW, 18). Using a similar argument, Pound cites with approval the economic and aesthetic analyses of Medieval Catholicism. He discovers dissociation at work in Dante's notion of usury and in the fact that the church determined the proper or acceptable level of usury. In *ABC of Economics,* he goes so far as to say "You can study economics almost entirely as dissociation of ideas" (SP, 243). In critiques of Eliot he simply takes for granted the equation of economics and the careful use of language, naming both dissociation.

In another oblique reflection on contemporary ignorance of the connections between economic and artistic practice, language and other exchanges, the penultimate review of *After Strange Gods,* titled "Ecclesiastical History (or the work always falls on papa)," offers the following remarks on the modern abrogation of careful argumentation as a symptom of ethical decadence, a repetition of his initial complaint against Eliot's own "modern philological lecture":

> 4. Concurrently [with the end of argumentation, the invocation of authority]: the decline of Christian ethics. *The middle ages distinguished between SHARING and USURY.* In correct theology, as Dante knew it, the usurer is damned with the sodomite. Usury with sodomy as "contrary to natural increase," contrary to the nature of live things (animal and vegetable) to multiply. The medieval trading companies, beginning mainly with question of ship's cargoes; *risk shared proportionally by all participants.* (NEW V:12, 272–73; emphasis added)

Usury prevents production and profitable exchanges of goods, just as authority and authorship, copyright and authorized interpretations, prevent the free exchange of words and ideas. In opposition to this, Pound advances, however metaphorically, the generative power of dissociation—of metaphor. Like money kept in circulation, the tradition of culture remains vital and increases only when shared and allowed free rein across various discourses and cultures.

Pound summarizes his various uses of "dissociation" in the crucial instigation which opens the first *New English Weekly* review, "Mr. Eliot's Mare's Nest." The following is a programmatic statement—or at least an extraordinarily suggestive instance—of Pound's dissociative reading procedure:

> In the "Ages of Faith," meaning the Ages of Christian faith, religion in the person of the Church concerned itself with ethics. It concerned itself specifically with economic discrimination.

It concerned itself with a root dissociation of two ideas which only the last filthy centuries have, to their damnation, lost. . . .

Creative investment, productive exchange, sharing the profits of shared risk, were considered good. Destructive parasitism was forbidden. . . .

The battle was won by greed. The language of religion became imprecise, just as the language of all forms of modern flim flam, including popular philological lectures, has become imprecise. (NEW IV:21, 500; emphasis added)

This treatment of dissociation has perhaps prepared us to recognize this (non- or anti-)method of reading as a sort of "counter-thesis to what Mr. Eliot's book claims to be" (unpublished review, p. 1). But Pound's identification of it as such is again secreted away in that unpublished version with so many of the other logical and explanatory statements withheld from the published reviews. The status of such buried explanations, especially one that explains so little, is at best problematic. By itself that kind of dissociation fails to organize the reading lists, quotations, and associations generated against *After Strange Gods*. It is not a thesis in the usual sense. By now one should not expect from Pound a counterthesis advanced through examples according to the standards set by the very essayists he derides. After all, Pound's play of interpretations, of elliptical dissociations of the words and texts Eliot addressed in his book, offers a reading that unreads Eliot's orthodoxy, which, as has been shown, Pound identifies with the imprecision and rigidity of modern language and culture.

More positively, Pound there begins to articulate his contrary notion of Kulchur as a "productive exchange," which he also names "The Cultural Heritage" and "increment of association." Early in the reviews, then, he begins generating the various metaphors he opposes to Eliot's organic figure of the orderly incorporation of texts into *the* tradition. But Pound's language seems to swerve out of control from the very beginning, for his point about usury is so elliptical as almost to disappear in allusion and association, requiring reference to *The Cantos* or other texts where his treatment of usury is more discursive. In this way, the dissociation of the proper *use*—or acceptable *usury*—of tradition both names and activates his reading procedure, which in the inflationary spiral of its own economy, multiplies interpretations and associations instead of defining a program. Pound's equation of money, language, and the other media of cultural exchange rests on the lexical dissociation of "usury" from "sharing," which is not explicitly elaborated in these reviews but is nevertheless crucial to them. Elsewhere, he submits the definition and etymology of "usury" to the sort of reading he gave Eliot's "orthodoxy," and he makes much clearer his abuse of accepted terminology than he did with his revision of Eliot's word. He forces "usury" to yield and bear unwarranted meanings, so that he can offer sharing as a corrective. "Sharing" and "usury" are not etymologically related. But Pound seems to call on etymology to examine the casual relationship of the near homonym "use/usury" in his numerous polemics against "destructive Parasitism" and his homilies about a proper (non-Eliotic) use of culture and criticism. Again, one cannot assume with confidence that his apparently strategic borrowings from other languages and discourses, however suggestive, are fully under his control, especially in the following definition, relying as it does on puns, translation, quotation, and other abusive figures:

> You can START savin' business by *STOPPING to use one word to mean two different things . . .* by seeing or *learning the DIFFERENCE between usura and partaggio,* i.e. between usury and sharing the results of association. (*Esquire* V:1, 195; emphasis added)

Pound's own words seem to undo his suggested reform: can he mean "stopping to use" in the sense of "not using"? So it would seem. This may be to make too much of a simple malapropism, but Pound does use at least one word in two senses there: "association." "Sharing the results of association," a concise definition of both "The Cultural Heritage" and his proposed activity of literary criticial exchanges, covers both the linguistic and psychological phenomena of thought-word "associations" as well as "association" in the sense of a group of persons with shared interests. Thus, not only had Pound to go through Italian to highlight and untangle a pun on use and usury; his translation equivocates on the rule it ostensibly establishes. At the same time, however, it instigates any number of questionable associations, including the alleged pun that would attribute a common etymology to the two words.

In the reviews of Eliot, Pound seems to welcome this condition of his terminology, perhaps so that he can slip back and forth between his favorite topics of economics and the general "Cultural Heritage," a category he makes more inclusive—or at least less exclusive—than Eliot's orthodoxy. For example, against modern definitions of the proper limitations of theology and literature, against charges of his own heresy, he offers this "ethical dissociation," allegedly borrowed from the Medieval church:

> The Source of value is the *CULTURAL HERITAGE*
>   The *aggregate* of all mechanical inventions and correlations, improved seed and *agricultural* methods, selected habits of civilized life, *the increment of association.* (NEW V:12, 273; emphasis added)

Typically, that correction—really a disruptive supplement—of Eliot's orthodox notion of cultural tradition compresses a host of metaphoric substitutions and puns: culture and agri-(aggre-)culture; culture as the aggregate of all the elements of both senses of "culture," civilized life and the cultivation of seeds/words. One can follow these figures through other of Pound's definitions of "Cultural Heritage" and "increment of association," which rather confusingly are sometimes two names for the same thing and sometimes general category and specific example. Several of the essays collected in *Impact* develop and interchange these metaphors. But these take one beyond the immediate concern of Pound's attempt to answer Eliot.

"Increment of association," borrowed from C. H. Douglas's notion of the incremental nature of historical progress, is as close as Pound will come to a refutation of Eliot's "orthodoxy."[37] Clearly, he was not of a mind to submit a simple definition of his own preferred form of orthodoxy, though his insistence seems to suggest that he has. Instead, he completes his dissociations and ends his reviews with the phrase and a few examples of "increment of association." If dissociation is a critical or analytical phase, its opposite side—but without dialectical resolution—might be "increment of association," the interpretations and innovations Pound hopes to stabilize and exchange under that name.[38] However, dissociations and the free-play of language Gourmont suggested under that category subvert Pound's apparent aim of proposing a set of standards. The final review of the series, "A Problem of (Specifically) Style," for example, as allusive and elliptical a piece as Pound ever wrote, refers again to the value of working together in literary and aesthetic "associations." "Increment of association" is offered as a basis for ethics, for the professional ethics Pound saw absent from the literary scene. Rejecting the ecclesiastical for a scientific model of argument, he suggests nevertheless that the search for truth supersedes its capture, and one must proceed not by either/or but by both/and; thus the quest advances not by refutation of another but by a troping, incremental reformulation of his own position in order to go beyond it. Pound does not want his position equated with the received position, nor can he accept the simple displacement of one truth by another. Discussion equates positions, while equations transform:

> Lacking an ethical base there is no argument against the perpetual (as I see it) infamy of dragging discussion continually onto the unessential, and continually away from the search after truth and knowledge. . . . In no science can truth go forward when men are more anxious to show up another man's minor error, or to prove his failure of foolproof formulation,

than to use his perception of truth (however fragmentary) for a greater
perception and for the formulation of valid equations. (NEW VI:6, 127)

Next Pound proposes his standards of evaluation, which are much less
explicit than Eliot's "imprecise" orthodoxy, especially since his notion
of the act of evaluation continually transforms the standard. He asks,
"Does the 'increment of association' mean anything in their philosophy?
Have they a philosophy? Does the 'heritage' figure in their computation?
Does the increment of associated machinery carry a meaning to their consciousness?" (NEW VI:6, 128). Is not culture, he asks, a transforming
machine as well as an organic thing in the process of natural change?
And aren't poets' and thinkers' acts (of reading, thinking) revolutions of
that machine?

Coming as it does at the end of the reviews, and as a statement of his
stylistic differences with Eliot—the titulary promise of "A Problem of
(Specifically) Style"—the exemplary association of scientists (and philosophers) working together incrementally stands as yet another uninterrogated metaphor for Pound's reading method that seems always on the
horizon, no matter how exhaustively he has stated his own preferred reading lists, his favorite modern interpretations, or deployed his dissociative
procedure against others. In Pound's hands method is metamorphic, not
only because he so often depends on metaphors.

The debate with Eliot over *After Strange Gods* contains some of Pound's
most suggestive remarks about the modern canon, about canonical criticism, yet it falls short of a statement that might turn his readings into a
usable Program. Certainly, Pound's evasion of closure and limitation begins strategically. But one cannot say with confidence that he rigorously
(in either the Nietzschean or the logical sense) questions or accepts the
"free"—or at least exponentially incremental—"associations" his readings exhibit and sanction. He simply does not sustain his reading—even
of "reading"—long enough to analyze or exploit the threat his habitual
substitution of metaphors poses to the dream of totalization (or totalitarianism) he entertained despite his antagonistic interventions into the orthodoxy of systematics—or the systematics of orthodoxy. When Pound
encounters the same terminological instability in his own text that he criticized in Eliot's, he fails to reflect on his own procedures, to exploit or
correct the figurative play and other linguistic aberrations enhanced by
his elliptical reading style. One suspects that his own ambivalence in this
regard motivates his habit of repeating and slightly modifying key definitions and snatches of favorite arguments and aphorisms in several es-
sayistic, epistolary, poetic, and other less generically pure formulations—

each of which was ostensibly to address its issue once and for all and with Pound's reputed concision. In this way he can evade the limitations of logical argument while gaining the apparent consistency of repetition and further explanation. The unpublished commentary which glosses the fragmentary reviews of *After Strange Gods* is a case in point. Fragmentary argument, even the fragment, exceeds the regulating law and cultivates the activity of interpretation.

Finally, one can easily redirect Pound's complaints against Eliot's interested misprisions back upon Pound's own practice of invoking rather than analyzing or systematically applying the methods and procedures of his own unorthodox interpreters, from Cavalcanti through Frobenius. Thus one grants a certain undeniable validity to Leavis's and Eliot's complaints against Pound's heterogeneity, his focus on individuals and individual words. Nevertheless, when the implications of some of Pound's dissociations are followed out, they would seem to account quite well for what might be called the expense of criticism, of accepting the task not of preserving but of propagating or cultivating an aesthetic heritage. To offer a *Guide to Kulchur,* he discovered, is to engage in a method that will not only transform the culture but will perforce demand its own modification. Method cultivates and clips itself, and the meanings or truths it promises to deliver are denied by the very procedure that makes the culture (as a quest) vital—that is, by criticism. As he ironically reflects on his own text that tentatively set forth the directions of cultural formulations, the guide will be misguided, but this is what keeps culture open:

> *Guide to Kulchur:* a mousing round for a word, for a shape, for an order, for a meaning, and last of all for a philosophy. . . . And they are still mousing around for a significance in the chaos. E.P., 20 June 1970
>
> (GK, Prefatory Note to the 3d edition [8])

CHAPTER FOUR

Writing "Frankly"
Gourmont and Pound on the Uses and Abuses of Science

Paris is the laboratory of ideas; it is there that poisons can be tested, and new modes of sanity be discovered. It is there that the antiseptic conditions of the laboratory exist. That is the function of Paris. It was particularly the function of De Gourmont.

POUND, 1915[1]

Conquérir l'Américain n'est pas sans doute votre seul but. Le but du *Mercure* a été de permettre à ceux qui en valent la peine d'écrire franchement ce qu'ils pensent—seul plaisir d'un écrivain. Cela doit aussi être le vôtre.

GOURMONT, 1915[2]

"Franchement d'écrire ce qu'on pense, seul plaisir d'un écrivain." "To put down one's thought frankly, a writer's one pleasure." That phrase was the center of Gourmont's position. . . . "Franchement, Frankly, is Frenchly," if one may drag in philology.

POUND, 1915[3]

Remy de Gourmont's precise influence on Pound's poetics, and thus on modernism, is open to speculation and likely to remain so, since Pound's exemplary Frenchman merely popularized unorthodox ideas, especially those of the Symbolists, Nietzsche, and several marginal scientists or enthusiasts of scientism. Many of the notions Pound attributed to Gourmont are little more than paraphrases and reading notes, drawn from other systems but not themselves comprising any system. In short, more reader than writer, he was, in Pound's terms, "an instigator," not a "master" or even an "inventor." This does not mean, however, that he failed to offer Pound a method; on the contrary, he did provide the poet a role model, not to mention a number of metaphors for assimilating literary traditions and scientific experiments into "modern culture." While literary critics and scientists have dismissed Gourmont as a dilettante, a sort of real-life Bouvard who collected, classified, and usually (con)fused already digested philosophy into questionable even literary notions, he was Pound's ideal man of letters, an inspired guide: "From Gourmont there proceeded a personal, living force. 'Force' is almost a misnomer; let us call it a personal light" (SP, 413). Pound was to invoke this "force" against the national, historical, and generic definitions of more than one literary orthodoxy.

Gourmont, whose only official jobs were assistant librarian at the Bibliothèque Nationale and literary editor of the newspaper *Mercure de France,* seems an unlikely culture hero. Pound acknowledges his passive and meditative character; nevertheless, he repeatedly praises the Frenchman as the ideal journalist for promoting young writers as well as the best ideas of the past. Moreover, somewhat ironically, Gourmont's celebrated dismissal from the library for publishing "Le Joujou Patriotisme," a parody of French nationalism and anti-German sentiment, became one of Pound's reasons for preferring "French thought" to Anglo-American poetry and German philosophy. Thus, as late as 1929, Pound stressed Gourmont's importance as a "type" and as a symbol for intellectual freedom: "We cannot afford to lose sight of his value, of his significance as a type, a man standing for freedom and honesty of thought. . . . How many well known and so called 'critical' writers pass their whole lives in, and how many entire periodicals are given over to, the production of statements agreeable to editors . . . but having nothing whatever to do with *thought, civilization,* or *honesty*"[4] (emphasis added). Thought, civilization, and honesty, all approbatory terms open to polemical interpretation, are the main Gourmontian qualities Pound emulated in his attempts to civilize literary America by reforming its periodicals and university curricula. We have already seen how he employed the rhetorical strategy of dissociation against the terms of Eliot's orthodoxy, but Gourmont's "thought" is equally instrumental in Pound's attempts to bridge science and poetry, America and Europe, if not to undo all the categories of literary history.

For example, while he was unable to cite either a literary or a scientific "masterpiece" to justify his claims for Gourmont, Pound erects this diffuseness of thought into a criterion of literary greatness, even a criticism of literary history, shared by the most important canonical poet-critics:

> The man was infused through his work. If you "hold a pistol to my head" and say: "Produce the masterpiece on which you base these preposterous claims for De Gourmont!" I might not be able to lay out an array of books equal to those of his elder friend, Anatole France. . . . You, on the other hand, would be in very much the same fix if you were commanded sud-

denly to produce the basis of your respect for De Quincey or Coleridge. (SP, 413)

Thus Gourmont's incorporation into the literary canon involves the questioning of canonization. Further, Pound argues, Gourmont's style, which is finer in the philosophical fragments than the poems, is not an end in itself, but a stimulant to (further) reading as well as a scientific—or poetic—revolution against philosophical abstraction: "Gourmont arouses the senses of the imagination, preparing the mind for receptivities. His wisdom, if not of the senses, is at any rate via the senses. We base our 'science' on perceptions, but our ethics have not yet attained this palpable basis" (LE, 345). Gourmont did not, in fact, offer an ethical system, or any other for that matter; instead, he interrogated systems. And, as Pound's metaphors suggest, he, like Nietzsche, practiced—and at the same time analyzed—a "physiological style," with which he criticized abstraction and assaulted complacent followers of the belletristic tradition.

This chapter will consider Pound's adoption of Gourmont's critical reading procedure, and hence the translation of Nietzsche's "transvaluation of values" and anti-Darwinism into a program of reading. In this regard, Gourmont's *Natural Philosophy of Love,* which Pound translated and to which he added an interesting afterword, is most important and demands close attention as a kind of theoretical text. But first we might consider Gourmont's ambiguous role in Pound's efforts to "conquer America." If Gourmont's transmission of specialized, particularly biological, knowledge into general culture is not always apparent in Pound's praise of the Frenchman, and if the revolutions each worked in literary history sometimes appear more nostalgic and reactionary than affirmative, this is perhaps because one views them from a certain American(ist) perspective. Pound and Eliot were not, however, the only modern American poets to debate the merits of Gourmont.

As London editor of *The Little Review,* foreign correspondent for *Poetry,* and Paris correspondent for *The Dial,* especially from 1915 to 1921, Pound transported the ideas and sometimes translated the works of Remy de Gourmont into American criticism. At the same time, he also practiced the journalistic ideal of allowing writers to "write *frankly,* " or as etymology allowed him to pun, "Frenchly," and thus to threaten American poetry. William Carlos Williams, who more than anyone else felt besieged by Pound, Eliot, and Gourmont, offers this fantastic commentary on a "reading scene" involving Pound, "the best enemy of United States verse," and his two cohorts:

Imagine an international congress of poets at Paris or Versailles, Remy de

Gourmont (now dead) presiding, poets all speaking five languages fluently. Ezra stands up to represent U.S. verse and De Gourmont sits down smiling. Ezra begins by reading "La Figlia che Piange." It would be a pretty pastime to gather into a mental basket the fruits of that reading from the minds of the ten Frenchmen present; their impression of the sort of United States that very fine flower was picked from.[5]

Here, in the "Prologue" to *Kora in Hell,* Williams finds Gourmont crucial to Pound's adoption of Eliot's Europeanism. Both Gourmont's traditional scholarship and his post-Symbolist verse threatened American independence and, even more seriously, the modernism of the two most influential American poets: "*Eliot's more exquisite work is rehash, repetition* in another way of Verlaine, Baudelaire, Maeterlinck—conscious or unconscious—just as there were Pound's early paraphrases from Yeats and his constant later cribbing from Renaissance, Provence, and the Modern French: *Men content with the connotations of their masters*" (Imag, 24; emphasis added). If Gourmont's exact role in effecting such "repetition" and "rehash" is not specified, Williams nevertheless suggests the Gourmontian reading course Pound followed in his early poetry and throughout the course of his critical writings: he borrowed a good deal of the Renaissance and Provence through *Le Latin Mystique* and *Dante, Beatrice et la poésie amoureuse,* and his "Symbolism" from *Le Livre des masques.*[6] For Williams, the idea of offering as the representative "American" poem a mood piece with an archaic title was as preposterous as it was Poundian.

Williams, however, fails to remark the differences between the two Gourmonts of his compatriots. Both Eliot and Pound acknowledged the influence and critical acumen of the Frenchman. For example, in *The Sacred Wood* he is "the most perfect critic" for his Benda-like refusal to commit treason against the apolitical posture of critics and academics (part of *La Trahison des clercs* was serialized in *The Criterion* in 1923). Similarly, Kenneth Burke calls Gourmont "an adept at pure literature" for his eclecticism and intellectual style.[7] But, for Pound, Gourmont was the champion of unpopular causes, a virtual anthology of "Mœurs contemporaines," more pedagogue than poet or critic.[8] Moreover, because he was both European and internationalist, Gourmont was subject to Pound's changing interpretations of several categories. Indeed, it was this flexibility in authorizing various positions that made Gourmont a particularly useful and long-lived figure for Pound's metamorphic *Kulchur.* Thus, as fellow worker, or as guiding spirit, in the Parisian "laboratory of ideas," Gourmont informed Pound's modernism as well as his antiquarianism—in ways Eliot disapproved and Williams did not (fore-)see. Years later, after reevaluating Pound's work and finding the language of *The Cantos* quite revolutionary, Williams dismissed Gourmont completely. That particular Frenchman fell from his consideration by 1920.

Nevertheless, Williams himself had recourse to European traditions to argue for his own revisionary history of modernism. By choosing figures from Spanish literature, which he identified as part of his own Mediterranean inheritance, he critically engaged literary history; like Pound, he had to become a re-reader before he could work a revolution in writing. Allowing some merit to Gourmont's cosmopolitanism, then, he marks Anglo-French modernism as a historical repetition, for example, of the way in which the Italian Renaissance had been reinscribed in Spanish verse, thus adulterating a native and natural poetry by infusing it with a high style. This double translation, as it were, which Pound, like Gourmont, tended to think of as an infusing of energy, troubled Williams, who suspected that the classical might be an enervation of native force:

> I do not overlook de Gourmont's plea for a meeting of nations, but I do believe when they meet Paris will be more than slightly abashed to find parodies of the middle ages, Dante and *langue d'oc* foisted upon it as the best in United States poetry. . . . *Ezra Pound is a Boscan who has met his Navagiero* [sic, Navagero]. (Imag, 26)

Williams's choice of literary precedents is part of his call for an "American idiom" and canon. In fact, the names Boscan and Navagero suggest his exemplary ur-American, Montezuma, whose quite literal *conquest* by Cortez prevented the development of a native American tradition. Juan Boscan, who conquered Spanish verse, and Hernando Cortez, the conquistador who made America Spanish, were contemporaries.[9] Boscan, the early sixteenth-century Catalan poet, brought the Italian Renaissance to Spain, when, under the influence and upon the direct suggestion of Andrea Navagero, the Spanish ambassador to Venice, he introduced sonnet conventions and the Italian verse form (*ottava real*) that became standard in Spain's national epics. These stylistic translations permanently transformed Spanish poetry, giving it a literary language distinct from ordinary Spanish, which was not suited to Italian rhythm and rhyme schemes. Williams elsewhere complains of this Italianate Spanish, just as he condemns Eliot's distortions of the American (as opposed to the English) language in, for example, "La Figlia Che Piange," with its Italian title and modified sonnet form. Therefore, in Williams's formulation, Eliot is the modern Navagero and Pound the modern Boscan, who introduced Eliot's imported verse forms in several American journals. But Gourmont, more for his criticism than for his poetry, also played the role of foreign agent, or Navagero, when, in the *Mercure de France,* he popularized both German philosophy and continental literature, hence influencing not only French but eventually American poetry. Moreover, as Boscan had transplanted the general ideals and attitudes of the Renaissance onto Spanish soil, by translating Castiglione's *Il Cortegiano,* Pound stressed the benefits of the Paris salons where Gourmont's works were discussed. Thus, if Williams's anecdote can bear the strain of our own rather far-fetched historicism, the conversation of Bembo and the other courtiers suggests the free exchange of ideas in which Gourmont, the reigning polymath, took part. In this way, Pound, whose translations of Gourmont's prose appeared in the same year (1919) as Williams's "Prologue," was both Boscan and Navagero, promoter of Eliot and ambassador to French culture. Gourmont, then, is an "instigator," one who merely sets the scene of conflict; he is one of a series of possible reference points in a chain of translations, not a source.

I have perhaps forced too much out of passing allusion, for the complexity of Williams's own historical revisionism remains to be elaborated, beginning with the radical prose of *Spring and All* (1922) and with the historicist rewriting of *In the American Grain* (1925), where he suggests that the literary and cultural implications of Cortez's conquest eventually contaminated the "origins" of an American native writing just as it undid the "ground" of a native American culture. If his polemics against Pound and Eliot, whom he later came to distinguish, seem to partake of a certain ethnocentric nostalgia, it would be wrong to dismiss Williams's remarks as simple xenophobia or nativism. He later identified the immediate cause of the "Prologue's" violent rejection of cosmopolitanism and tradition; it was the publication of "The Love Song of J. Alfred Prufrock" which Williams took to be the first salvo in the battle for influence over the future, as well as an objectionable reading of the past:

> When I was halfway through the Prologue "Prufrock" appeared. I had a violent feeling that Eliot had betrayed what I believed in. He was looking backward; I was looking forward. He was a conformist, with wit, learning which I did not possess. He knew French, Latin, Arabic, god knows what. I was interested in that. But I felt he had rejected America and I refused to be rejected and so my reaction was violent. I realized the responsibility I must accept. I knew he would influence all subsequent American poets and take them out of my sphere. I had envisioned a new form of poetic composition, a form of the future.[10]

Certainly, Williams's own form of the future incorporates the experiments of contemporary European visual artists and poets, and especially the transplanted artists of the New York Cubist, Dadaist, and Futurist movements, as well as figures from his version of the countertradition. In this, he approaches Pound's (Gourmontian) endorsement of modern experiment and historical heresies. Williams never came around to accepting Eliot, but his epic, *Paterson,* shares the "compendium"—if not the compendious—form of *The Cantos,* which clearly draws upon general culture and the various discourses that were Gourmont's special province. And while we cannot here follow out the parallel yet intersecting lines of the Pound-Williams relationship, Williams was not able to dismiss Pound's "repetitions" and "connotations" which he came to praise, if not in these terms, as part of *The Cantos'* revolution in American language and poetry.[11] Indeed, his rejection of Pound-Eliot-Gourmont, and all he made them represent, would seem to confirm Paul de Man's observation about the differential repetitions of "modernity" throughout literary history:

> Modernity turns out to be indeed one of the concepts by means of which the distinctive nature of literature can be revealed in all its intricacy. No wonder it had to become a central issue in critical discussions and a source of torment to writers who have to confront it as a challenge to their vocation. They can neither accept nor reject it with a good conscience. When they assert their own modernity, they are bound to discover their dependence on similar assertions made by their literary predecessors, their claim to being a new beginning turns out to be a claim that has always already been made.[12]

By this inexorable logic, the louder Williams proclaimed his rejection of the European past, the more deeply he engaged the questions of reading and literary history Pound attributed to Gourmont.

Nevertheless, Williams had good reason to feel bombarded by Pound's enthusiasm for Gourmont and to suspect that the Frenchman was displacing American poets—both from little magazines and from the canon. For example, from 1919 to 1921 nearly all the American journals seemed devoted to Pound's version of Gourmont. In 1919, Pound edited a special double issue of *The Little Review* devoted to analyses of Gourmont's criticism and poetry; these included his own long essay, "Remy de Gourmont: A Distinction," and critical tributes by Frederick Manning, Richard Aldington, and John Rodker.[13] Rodker's essay, "Gourmont—Yank," is interesting because it praises Gourmont's transformations of any canon claiming to be modern or American. He argues that Gourmont simply repeats Poe and hence becomes part of a rather eccentric American canon, with Poe as representative poet-critic. Rodker's claim that Gourmont is American was justified by Poe's critical writings, which have always been more well received in France than in America. But he characterizes this cross-cultural exchange as a transformative reading, as the very sort of mixture Pound achieved by interfiling a heterogeneous poetry and commentary: "I suspect that part of De Gourmont's 'mission' was to fill in the holes left by Poe. Certainly they illuminate each other, are commentary exegesis."[14] This reading of Gourmont's "reading" revises Franco-American letters in such a way that the two become nearly indistinct, but French culture nevertheless triumphs over American backwardness, in a displacement Williams would have found particularly irksome: "Nor has De Gourmont ever surpassed the description of the interior of the young man's palace in 'The Assignation,' (the Rimbaud-like purity . . .) and how much more remarkable Poe's preoccupation [with stylistic purity] seems when it is remembered that it took place in a barbarous country— one to which Whitman was infinitely more suited."[15] By this account, Poe gains stature for contributing to Symbolism and to Gourmont's stories; that is, American writing becomes literature only by passing through France.

It might be said that Williams has his Whitmanesque revenge against this sort of Francophilia when he says: "One day Ezra and I were walking down a back lane in Wyncote. I contended for bread, he for caviar. I became hot. He, with fine discretion, exclaimed 'Let us drop it. We will never agree, or come to an agreement.' He spoke then like a Frenchman, which is one who discerns" (Imag, 26). But Pound had something more in mind than discernment—more, that is, than the French intellectual's reputedly over-refined taste. Moreover, Pound's endorsement of Gourmont's (French) open-mindedness and antinationalism appears in the text of the "Prologue," not only in Williams's dismissive fable, but also in the long quotation from Pound's letter about what it means to be a "REAL American." In this letter of 10 November 1917, Pound cites Williams's own foreignness by way of affirming the inescapable yet salutary foreign influences on American verse:

> What the h—l do you a blooming foreigner know about the place. Your *père* only penetrated the edge and you've never been west of Upper Darby, or the Maunchunk switchback.
>
> Would H. [Pound's original letter specifies "Harriet," as in Monroe] with the swirl of the prairie wind in her underwear, or the Virile Sandburg recognize you, an effete easterner as a REAL American? INCONCEIV-ABLE!!! . . .
>
> You thank your bloomin gawd you've got enough Spanish blood to muddy up your mind, and prevent the current American ideation from going through it like a blighted colander. (Imag, 11)

Here, Pound's definition of "REAL American" cuts both ways. On the one hand, he stresses the importance of rootedness, of being born in America and knowing its frontiers; on the other hand, he insists that American thought needs the infusion of Europe's more substantial tradition; even if this tradition is "muddy," it nevertheless prevents the usual American habit of passively relaying popular ideas—the colander mode of ideation Williams escapes. Thus, if he does not go so far as to "dissociate" two notions of "American" or two ways of relating to "the foreign," Pound does suggest that America is a place of transformation, where ideas drawn from Europe, perhaps more through a process of condensation or amalgamation than simple translation, become original and native—again. Even as he had claimed to advance beyond Whitman's influence, to escape his own American inheritance, he muddies his own poetic genealogy. For instance, in "What I Feel About Walt Whitman," a short essay published eight years before he corrected Williams on the point of American legitimacy, he confesses his ambivalence about his native poetry:

> Personally, I might be very glad to conceal my relationship to my spiritual father and brag about my more congenial ancestry—Dante, Shakespeare, Theocritus, Villon, but the descent is a bit difficult to establish. *And, to be frank, Whitman is to my fatherland (Patriam quam odi et amo* for no uncertain reasons) *what Dante is to Italy.* (SP, 145–46; emphasis added)

Thus Pound outlines a reading program which virtually promises to find Whitman reinscribed in Dante. That is, American writers should (re-)turn to Europe in order to go beyond it, though they will never escape being American. And while it is perhaps going too far to say that his "frank" feelings about Whitman owe anything to Gourmont, *Le Latin mystique,* which finds Symbolism in Provençal texts, in the same sense that he proposes uncovering Whitman in Dante, was one of his first European "textbooks," providing him "new" ideas and also an example of how to assimilate foreign and ancient texts—how, that is, "to make it new."

In this vein, in the part of the letter to Williams which does not appear in the "Prologue," he recommends: "Your sap is interrupted. Try De Gourmont's 'Epilogues' ('95–'98). And don't expect the world to revolve around Rutherford" (L, 123). The part of the letter Williams does quote includes a long passage from *Epilogues,* which Pound introduced with a brief (self-)reflection on an American writing he defines as self-consciously appropriating a textual tradition: "Let me engage in the American habit of quotation." And here I quote Williams's quotation at some length because it indicates the cosmopolitanism prerequisite to Pound's own reading procedure:

Si le cosmopolitisme littéraire gagnait encore et qu'il réussit à éteindre ce que les différences de race ont allumé de haine de sang parmi les hommes, j'y verrais un gain pour la civilisation et pour l'humanité tout entière. . . .
[Pound's ellipsis]
    L'amour excessif d'une patrie a pour immédiat corollaire l'horreur des patries étrangères. . . .

Cette folie gagne certains littérateurs et le même professeur, en sortant d'expliquer le Cid ou Don Juan, rédige de gracieuses injures contre Ibsen et l'influence, hélas, trop illusoire, de son oeuvre, pourtant toute de lumière et de beauté. (Imag, 12)

This passage, quoted by Pound and reinscribed by Williams, is from Gourmont, who argues that the raison d'être of literary cosmopolitanism is the abolition of "differences" that lead to racial, or "blood," hatred, and thus to war. This is in keeping with Gourmont's constant plea for the pacification of Europe. Yet, Chapter Three's examination of "dissociation" showed, Gourmont—and Pound—respected stylistic and cultural differences in a cross-epochal *cosmopolitanisme*—which is, after all, a gathering and not a synthesis. Gourmont also suggests the terms of Pound's reformations of the American literati and curricula: that explicating, or advocating, of the great works of other traditions should not involve the dismissal of one's own modernism; and, conversely, that one's modernism and Americanism do not demand wholesale rejection of European traditions. In this way, "Pound's Gourmont" argues equally against Eliot's continental nostalgia and Williams's nativist iconoclasm, upholding the ideals of heterogeneity and reading against both writers.

Gourmont raised the question of national characters with regard to the pedagogic role of journalists, in a way that Pound imitated. In an essay about Herbert Spencer, whom he claims was better understood by the French than the English, Gourmont explains the goal of writing "frankly," in the way Pound opposed the term to Williams's usage. Gourmont's nationalism involves a certain self-effacing irony when he opposes the French habit of playing with ideas to the English (and American?) habit of banishing them:

C'est au journaliste à faire son éducation intellectuelle. En France, le journaliste n'y manque pas. *Soyons juste: pour un grand écrivain qu'il méconnaît ou qu'il bafoue, il en célèbre dix avec abondance et avec enthousiasme.*
. . .

Si le journal anglais est une école d'ignorance, le journal français est une *école de badauderie.* (emphasis added)[16]

Pound would say that if the French journalist praised ten for every one great writer he knew well enough to criticize, this was still preferable to

the English journalistic practice of refusing to entertain new ideas. Indeed, Pound left London, his first school of European culture, because
he found it, in Gourmont's phrase, a "school for ignorance" where one's
efforts to publish new ideas were constantly frustrated. But "to be just,"
as Gourmont says, the same charges of uninformed enthusiasm could be
brought against Pound because he seems everywhere to prefer "gists and
piths" to careful analyses. More than once he praises Gourmont for keeping ideas at a distance, for playing the part of a *bricoleur,* if not a *ba-
daudier* (star-gazer, idler)—that is, a sort of jack-of-all-trades, one who
collects fragments rather than advances logical arguments, and thence
tests ideas in the improvised "French laboratory."[17]

Just so, Pound's inaugural and most widely publicized translation of
Gourmont before *The Natural Philosophy of Love* was "Dust for Spar-
rows," a sheaf of posthumous notes published serially over nine issues
of *The Dial* (September 1920 to May 1921).[18] In his "Translator's Introduction," Pound insists on both the literary and pedagogic utility of the
paragraph-length fragments that abbreviate Gourmont's notions, ranging
from insect behavior to architecture, philology to telepathy. Recalling the
letter in which Gourmont had proposed "*civiliser l'Amérique*" or "*conquérir l'Américain,*" he says: "Because of ill health he could send me,
for a proposed literary venture, only 'indications' of ideas, not '*pages
accomplies.*' The section of his unpublished work headed Dust for Spar-
rows (*Poudre aux moineaux*) is presumably of that period." Pound goes
on to characterize the fragments as the movements toward, even the spaces
between, thoughts: "The following pages are not intended as epigrams;
they are indications and transitions of thought."[19] Subtly, then, the themes
of Pound's "reading"—his reading-translation-metaphor—come into play
when he finds the crux of Gourmont's *thought,* which is at once critical
and organic, exemplified in one of the fragments: "Every thought is a
stem, potentially flower and fruit; some suggest, *question the unknown,
interview truth;* some affirm [All this little series of paragraphs should
be taken in relation to Gourmont's essay on style. E.P.]."[20]

One does not know whether Pound refers to one of Gourmont's many
essays specifically about style (*Le Problème du style* and "Du Style ou
de l'écriture" are likely candidates) or to *The Natural Philosophy of Love*
(*Physique de l'amour*), which he took to be representative of Gourmont's
style of thought. In either case, Pound treats all of Gourmont's works as
a compendium of poetic yet scientific notions which can be used to "interview truth," to question scientific—and poetic—method. Despite the
methodologically suspect procedure of drawing metaphors, or even metaphors of metaphors, from fragments, Pound recognizes the critical thrust

of Gourmont's "sensuous wisdom," his "heterogeneous works." Therefore, several years before undertaking to translate *Natural Philosophy,*
Pound ascribes a unity or authorial self to the contradictory mixture of
feeling and thought, poetry and science, physiology and metaphysics:

> He recognizes the right of individuals to *feel* differently. Confucian, Epi-
> curean, a considerer and entertainer of ideas, this complicated sensuous
> wisdom is almost the one ubiquitous element, the "self" which keeps his
> superficially heterogeneous work vaguely "unified."
>
> The study of emotion does not follow a set chronological arc: it extends
> from the *Physique de l'Amour* to *Le Latin Mystique;* from the condensation
> of Fabre's knowledge of insects to
>
> "Amas ut facias pulchram"
>
> in the Sequaire of Goddeschalk (*Le Latin Mystique*). (LE, 340–41)[21]

Here Pound suggests the wider methodological implications of Gourmont's procedure: the revitalization of abstract by poetic language. In
addition to using Gourmont's works and *position*—both his "views" and
his "function" as storehouse (way-station?) of European ideas—against
the national and generic definitions imposed upon literature, Pound adopts
Gourmont's (Nietzschean) project of reversing the *entropy* of Western
philosophy—that is, the exhaustion of its creativity and its rejection of
linguistic *tropes*.[22] If Pound adopts the very tropes by which Gourmont
hoped to *turn* scientific and philosophical thought back toward poetry's
original (figurative) force, this too is in keeping with the Frenchman's
style—and with what Derrida has described as the inevitability of treating
metaphor *more metaphorico*. Thus, in a passage which draws upon what
Pound took to be the central notion of *Natural Philosophy,* "spermatic
thought," he at once dismisses philosophical and psychoanalytical (German) thought and yet finds system (even an Aesthetics) in Gourmont's
treatment of insect sexuality:

> From the studies of insects to Christine evoked from the thoughts of Diomede,
> sex is not a monstrosity or an exclusively German study. . . . Sex, in so
> far as it is not a purely physiological reproductive mechanism, lies in the
> domain of aesthetics, the junction of tactile and magnetic senses. (LE, 341)

In passing on to *The Natural Philosophy of Love* we might note that both
the notions of sensuous wisdom and the endeavor to make philosophy
sensuous—or, more appropriately, sensual—brings Pound, by way of
Gourmont, into relation with Nietzsche, that German philosopher who
asked: "*What is truth?* Inertia; that hypothesis which gives rise to con-
tentment; smallest expenditure of physical force, etc." (WP, 291).

II

His translation of *Physique de l'amour* brought Pound as close to Nietzsche as any work he undertook.[23] Yet he hardly came into direct contact with Nietzsche's ideas, which came to him through more layers of interpretation and translation than had been inscribed in the originals. According to his own much quoted observation, Pound knew the German philosopher through the distancing—or aestheticizing—medium of Gourmont's French thought: "Nietzsche has done no harm in France because France has understood that thought can exist apart from action" (SP, 421). Written about Gourmont during World War I, at the moment when Nietzsche's *reichsdeutsch* interpreters were turning such texts as *Will to Power* into endorsements of the German *Kultur* Nietzsche had so violently criticized, Pound's apology echoes the typical English renunciation of the once fashionable moral and aesthetic philosopher. Yet Pound also raises questions about the translation and transformation of German philosophy into other discourses and through several national cultures; which or *whose* Nietzsche did Pound inculcate? Is it possible that, despite a certain French indirection, the American poet could have practiced such a rigorous critical program as "transvaluation of values"? I will argue that by adopting "Gourmont's Nietzsche," he does indeed become affiliated with the rhetorical critique of philosophical and scientific values.

The Natural Philosophy of Love does not, however, present "pure" Nietzsche. Instead, Nietzsche and Herbert Spencer, Gourmont's ideal critics of philosophy and science, respectively, preside over the Frenchman's critique of the moral and methodological prejudices of German philosophy and English naturalism. In a sort of *reductio ad absurdum* of taxonomical rigor, Gourmont classifies and details the sexual excesses of insects in order to expose the moralistic anthropomorphism that finds religious and social conventions in the animal world and thereby distorts empirical or scientific investigation. At the same time, he questions the mutual exclusivity of the poetic and scientific discourses, noting that both rely on the same rhetorical figures and strategies—usually the uninterrogated "metaphors," as he terms them, of "God," "Country," "Family," and the like. Further, whether or not Pound recognized the Nietzschean character of his own "literary" application of Gourmont's "science textbook," his "Translator's Postscript" addresses fundamental issues of both literary and scientific method, ending in a "transvaluation" of the "traditional wisdom" about modern readers and writers and the tradition. In this way, Gourmont—like Fenollosa, Frobenius, and Agassiz—involves Pound in the scientific revolution effected by Darwin and analyzed—or, better, opened to interpretation—by Nietzsche and later by the American (Henry) Adams.[24]

Ironically, Nietzsche seems to have predicted his own fate at the hands of such translators and popularizers as Gourmont. This is not to say that he posthumously endorses a *proper* Nietzsche. Quite the contrary, he delights in the inevitable distortions of the journalistic and pedagogic transmission of his inherently disseminative thought; just as he undid the orderly (genealogical) succession of philosophical ideas, his own texts refuse to generate system or to stay within one language. Nevertheless, he expresses a taste for the playfulness, even the rhetorical self-consciousness, of French interpreters of Wagner and German philosophy—and, by implication, his own works. For example, one of the lighter sections of *Will to Power* begins to elaborate this running multiethnic joke, which might be thought to parallel Gourmont's—if not Pound's —observations about French and English readers:

> Toward a characterization of national genius in relation to what is foreign and borrowed:
>> The *English* genius coarsens and makes natural everything it takes up;
>> the *French* makes thin, simplifies, logicizes, adorns;
>> the *German* confuses, compromises, confounds, and moralizes. (WP, 438)

This schematic can be filled in with the names of the "Nietzscheans" Pound knew: The "Englishman" Wyndham Lewis "coarsened" *Übermensch* into his amoral Machiavellian hero; Frau Förster-Nietzsche "moralized" and certainly "compromised" her brother to the *Reich;* the French Symbolists, who inspired A. R. Orage's "Dionysian Nietzsche," "thinned" the early, Wagnerian Nietzsche into an "aesthete," a Symbolist as it were. In France, then, Nietzsche was treated as a poet—and usually by poets.

In a lengthy obituary, "La Morte de Nietzsche," Gourmont comments directly, if moralistically, on the French reception of Nietzsche. He complains that while philosophers (the French Academy) readily accepted Nietzsche as "the great poet," they ignored his role as "creator of values" ("professeurs de philosophie . . . acceptant l'écrivain, proclamant le grand poète, ils dédaignent le créateur de valeurs").[25] For his part, Gourmont praises, but refuses to separate, both the aesthetic and the critical aspects of Nietzsche's text, claiming that, out of entirely new and personal ideas (*"des idées personnelles et toutes neuves"*), he created a procedure for interrogating all Idea. Thus, Gourmont was in 1900 already a rereader of Nietzsche. Yet his notion of a comprehensive but minutely analytic and individualistic Nietzsche does not, in fact, differ greatly from what Nietzsche himself expected of French habits of reading and rhetoric.

If, as noted above, Nietzsche sometimes dismisses French interpreters, in the "Why I Am So Clever" section of *Ecce Homo,* he expresses high regard for their refusal of "high [German] seriousness." For example, he prefers Baudelaire's interpretation of Wagner to that of the "Wagnerites": "Who was the first *intelligent* adherent of Wagner anywhere? Charles Baudelaire" (EH, 248). This is not to say that Baudelaire had, like the Germans, interpreted Wagner's philosophical (Hegelian) importance or placed him in the great epic tradition. Rather, the French *décadent* saw Wagner as the reflection of his own artist's persona, his modernism. Thus, Baudelaire was also "the first to understand Delacroix—that typical decadent in whom a whole tribe of artists recognized themselves" (EH, 248). Here Nietzsche shifts the burden of interpretation, if you will, to artists, thereby exchanging the creative for the critical discourse and disturbing the easy distinction between the two. By the time Gourmont wrote *Le Livre des masques* (1896), a series of critical biographies of the Symbolists, Nietzsche was seen to authorize the cults of the individual and the artist: "The capital crime for a writer is conformism, imitativeness, submission to rules and teachings. The work of the writer must be not only the reflection, but the enlarged reflection of his personality" (SW, 181). This naive revolutionary stance—really a caricature of Nietzsche— gains interest because Gourmont uses it in his own battle against systematic philosophy: when, for example, he attacks Kant, his most frequent target, for rejecting artistic and "scientific" freedom and choosing instead "to throw himself to the aid of a shipwrecked morality" (SW, 181). This complaint against Kant's moral and religious conformism is thematized in *The Natural Philosophy of Love*.

Nietzsche stressed the essential foreignness (the oxymoron is appropriate) of art and artists, strategically opposing Wagner's—and his own— Parisian sensibility to German virtues: "The way I am, so alien in my deepest instincts to everything German. . . . [T]he first contact with Wagner was also the first deep breath of my life. . . . I revered him as a *foreign* land, as an antithesis, as an incarnate protest against all 'German virtues'" (EH, 247). Here Nietzsche does not simply replace one nationalism with another but makes such definitions absurd by ironizing the role of "the enemy" of the German state, France. If his artist-decadent has a home, it is Paris, which as Pound said in his own argument against (American) provincialism, is "the capital of Europe." Nietzsche attributes to "Parisian seriousness" the very sort of psychological subtlety, the dramatic and ironic skill he demanded of his "new philosopher":

> As an *artist* one has no home in Europe, except Paris: the *délicatesse* in all five artistic senses . . . the fingers for *nuances,* the psychological mor-

bidity are found only in Paris. Nowhere else does one have this passion in questions of form, this seriousness in *mise en scène*—which is Parisian seriousness *par excellence*. (EH, 248)

Here one can detect the symptoms of Gourmont's use of Nietzsche, from whom he claimed to have learned the skill of examining the forms or philosophical Categories, even the connotations of the individual words of philosophical argument. Moreover, Nietzsche was featured prominently, if sometimes anonymously, in his dramatizations (*mise en scène*) of his methodological battles against academic science, philosophy, and the other entrenched systems. It is, then, in his idiosyncratic, and not entirely *French,*—or for that matter, frank or open—readings of Nietzsche as the rhetorician who embarrassed metaphysics that Gourmont approaches the (de Manian) *rigor* of the Nietzschean critique.

In a remarkable pasage which presages the "New [French] Nietzsche," Gourmont credits "the creator of values" with teaching philosophers how to "deconstruct" metaphysics. While we cannot here explore Gourmont's place in (the margins of) the Nietzsche-Heidegger-Derrida critical enterprise, it is noteworthy that he not only deploys deconstructive strategies (irony, "poetic" or figurative language, the enigmatic notions or "undecidables" that upset discursive norms), but also uses the figure of "the house of philosophy" destroyed through the critique of its grounding concepts to characterize Nietzsche's contribution. In this way, he "reads," as it were, before the letter, Heidegger's nostalgic (even "onto-theo-logic") metaphor of "Building Dwelling Thinking," the poeticizing (re-)construction of *authentic* thinking that survives the death of metaphysics.[26] Whether or not Heidegger can be implicated here, Gourmont, who adopted the pose of destroyer of metaphysics and champion of poetic against abstract language, had as his adversaries those "eternal"—canonical?—philosophers ("*les philosophes, les eternals professeurs*") who failed to recognize Nietzsche's overthrow of metaphysics:

> Je crois qu'il y a un grand changement. Nous avons appris par Nietzsche à déconstruire les anciennes métaphysiques édifiées sur la base de l'abstraction. Chacune des antiques pierres d'angle, les voilà poussière, et toute la maison s'est écroulée. Qu'est-ce que la liberté? Un mot. Alors, plus de morale, sinon esthétique ou sociale.[27]

This deconstructive or deconstructing Nietzsche is instrumental in Gourmont's critique of the moral prejudices of modern science, and we will encounter him again in the text of *The Natural Philosophy of Love*. It is, however, in an essay about Herbert Spencer, whom Gourmont took to be a subtle rhetorician and, not without contradiction, the founder of at least one system, that he explains the ideals and procedures of the deconstruction of scientific method. Perhaps not accidentally, the "house of philosophy," as opposed to museums—of science, industry, and the fine arts—also figures in his tribute. Therefore, we can best approach Gourmont's Nietzsche by way of a detour through Spencer's fanciful "Synthetic Philosophy" and his "sociology."

Spencer participated in the late-nineteenth-century quest for an "umbrella" discipline that would cover all the arts and sciences and account for the "life of man." To this end, he projected a ten-volume work which would interpret the progress and progression of human societies and accomplishments on the model of (his revised version of) Darwinian evolution. "Synthetic Philosophy," which names both his unfinished *magnum opus* and the "science" it inaugurated, catalogues the first principles and key discoveries of such disciplines as psychology, biology, anthropology, and the arts; it remains throughout more concerned with assembling observations from—rather than producing analyses of—previous investigations in the various discourses. Spencer's principal qualifications were that of journalist, free-lance "expert," and unemployed engineer; indeed, had he been in Paris, he might have been an intellectual, a vocation unknown in England and America. In other words, he did not master one discourse; rather, he challenged the assumptions and mutual exclusivity of several. Though Gourmont's treatment of him as a critic of science is rather idealized, Spencer's search for method would seem to parallel the efforts of Gourmont and Pound on behalf of science and poetry—and of a "science of poetry."

Spencer's principal contribution and recurrent theme is the notion of "compound evolution" which, by way of revising Darwin's monolinear and progressive model of natural selection, identifies two processes, integration and differentiation. The latter is the source of change and the privileged category that issues in the individual, which Spencer places above the species and, in the case of humans, above society. Differentiation also accounts for the sudden, even violent, transformations effected by artistic genius. These Darwin failed to address, for the slow process of natural selection favors neither the existence nor the persistence of individual—which is to say, "monstrous" or "unnatural"—differences issuing in abrupt changes. Both Spencer's biology and Nietzsche's aesthetics agitate the Darwinian model and contribute to Gourmont's accounts of aberrations and interruptions in the evolution of insects. In this "Translator's Postscript" Pound turns these theories back into questions of aesthetics, though he uses his "artistic instinct" rather than citing any sources:

I believe, and on no better ground than that of a sudden emotion, that the change of species is not a slow matter, managed by cross-breeding, of nature's leporides and bardots. I believe that the species changes as suddenly as a man makes a song or a poem, or as suddenly as he *starts* making them, more suddenly than he can cut a statue in stone, at most as slowly as a locust or a long-tailed Sirmione false mosquito emerges from its outgrown skin. (NPL, 212)

Here Pound employs his own comprehensive science, a mixture of biologism, psychology, and aesthetics loosely held together by analogies. Not without contradiction, he advances this "unified" model to explain the individual genius, who, he punningly suggests, "starts" new movements and advances culture by fits and "starts"—that is, discontinuously.

Similarly, Spencer's project of charting human accomplishments rested on one principle, evolution, and assumed the perfectability and the ultimate "integration" of Western civilization into a "great society"—one which was to resemble that of Spencer's Victorian England. Yet at the same time he focused on heterogeneous details and unaccountable artistic and scientific inventions; that is, on the differential character of compound evolution. He thus disturbed the Darwinian model only to raise procedural and methodological problems which he failed to resolve. Gourmont notes his inconsistencies, suggesting that, however chimerical, these stimulate thought and derive from Spencer's wish to prevent evolution from *devolving* into fairytale:

> L'idée d'évolution n'est pas une chimère, quand on la conçoit purement mécanique, c'est-à-dire quand on se garde de lui donner les buts moraux ou bienfaisants. Spencer n'eut pas cette prudence. Il crut que l'évolution évoluait vers le bien, vers le bonheur: et revoilà les contes à endormir les petits enfants.[28]

Gourmont's disdain for the "watering [or wearing] down" of science can be linked to the nineteenth-century anxiety that human creativity— indeed, the whole physical universe—was becoming entropic. The struggle of art against convention, science, and the mechanical, a theme shared by Nietzsche, Spencer, Gourmont, and Pound, was one way of reversing this process and escaping the laws of thermodynamics. In Gourmont's case, the desire at once to glorify science and philosophy and to save them from themselves issues in parodies of, even animal fables about, the great scientists and rationalists (Gourmont's/Pound's term "ignorantist" denotes the rationalist or idealist tenet that experience will not yield knowledge or Idea). For example, in *The Natural Philosophy of Love,* Kant not only rides a child's "hobby-horse" or trained pony; he is one:

> Kantian ignorantism is the masterpiece of these [philosophical] training exercises, where, starting from the categoric stable the learned quadruped necessarily thither returns, having jumped through all the paper disks of scholastic reasoning. Observers of animal habits fall regularly into the prejudice of attributing, regularly, to beasts directive principles which only a long philosophic education and especially Christianity have rammed into restive human docility. (NPL, 87)

This whimsical account of philosophical misprision is also a warning against the loss of creativity and the creative impulse which war against docility—against system.

Spencer became Gourmont's hero, his champion against Kant, Darwin, and the bourgeoisie, by virtue of a certain anarchical genius—the same sort of knack for collecting unaccountable data that Pound attributed to Gourmont. For example, while acknowledging that Spencer's critical sense ("*sens critique*") was mediocre ("*plus médiocres*"), he adds: "Mais son intelligence était vaste. C'était un immense palais sans art et sans gout ou il entassait les faits dans les armoires; de temps en temps, il en dressait l'inventaire. C'est de qu'il appelait faire la synthese des connaissances humaines."[29] Thus, Spencer's "thought," if not his "Synthetic Philosophy," is a great baroque palace—or, better, that favorite nineteenth-century *space,* a "baroque gallery" or artist's studio in which were displayed representative works of any number of arts and epochs: a reproduction of a Michelangelo statue here, a bust of Caesar there, a Velázquez over there, a large closet—or armoir—with every manner of scientific or surgical implement over in the corner. Note that Spencer's "palace" lacks art and taste; his armoires contain an inventory of heterogeneous facts capable of many different arrangements and interpretations. Therein lies his utility for Gourmont, who always had an eye out for the odd fact, the pretentious citation. Out of this anarchy of details Gourmont fashioned his own rhetorical strategy of dissociating layers of interpretation and questioning whatever ideas remained.

For Pound, Gourmont's own writings were storehouses or palaces of ideas, though, of course, he never questioned his hero's critical sense, in matters of selection or of method. Curiously, this figure of a palace features in Pound's own aesthetic even as a rhetorical strategy. For example, *The Tempio,* Malatesta's palace-temple-collection, housed what Pound claimed were the finest remains of pagan along with some of the most interesting curios of Christian art. Moreover, Pound's favorite *condottieri* brought together the best artists of several media and styles to erect his monument and create the treasures it housed, including the corpses of its architects. A passage from Canto IX can be taken as a figure for

Pound's own efforts to gather the salvageable fragments of civilization into his epic and his reading text. *The Tempio,* which has been thought as tasteless as the *Hearst Castle* at San Simeon, is part of his argument for a cultural synthesis: "[Malatesta] 'built a temple so full of pagan works'/ i.e. Sigismund/and in the style 'Past ruin'd Latium'/The filigree hiding the gothic/*with a touch of rhetoric in the whole*" (IX: 41; emphasis added). One does not know which sense of "rhetoric" Pound had in mind. Surely, "filigree hiding the gothic," which is to say, filigree upon filigree, suggests the sort of "ornamental metaphor" Pound everywhere rejects. However, since *The Tempio* recurs in his readings of papal corruption and the church's destruction of Classical beauty by a misplaced orthodoxy, Malatesta's masterpiece is also an interpretive metaphor. Hence this monstrous creation, "both an apex and in verbal sense a monumental failure" (GK, 159), becomes, like Gourmont's Spencer, a tool against aesthetic and other orthodoxies, as well as a figure of his own heterogeneous style.

Gourmont is more explicit about Spencer's role in reversing the essentializing tendencies of modern science and philosophy. He is a sort of antidote (*préservatif*) against both evil or unpropitious thoughts (*mauvaises pensées*) and rationalism (*rationalisme*)—the sort of "new mode of sanity" which arose when Gourmont tested for "poisons" in the "laboratory of ideas."[30] In his account, the English naturalist was part of a countertradition of experimental scientists unspoiled by System and its prejudices: "Spencer était naturiste, comme Bacon, et comme lui, collectionneur. Oubliez le catalogue et regardez dans les vitrines: c'est un beau musée."[31] While Gourmont might seem to legitimize Spencer by equating him with Bacon, the titular founder of the scientific method, he stresses those qualities anathema to method; his Bacon thereby disturbs science at its (non-)origin, by an invention that is as radical as what Thomas Kuhn has called the "change of paradigm." Gourmont again accomplishes this reading by using the figure of the collection, this time a museum full of exhibit cases that should be consulted intead of an exhibit catalogue, itself a questionable scientific text. There is never the suggestion of abandoning text for direct treatment of data, nor is there a *book* involved—rather a catalogue, an open text, another compendium. Thus, Spencer provides neither system nor unmediated experience, but an aleatory collection of museum pieces.[32]

One does not know whether Pound had this passage in mind when he used the figure of the museum to characterize his own reading program. And surely a chain of texts comprised of temples, museums, and the "thought(s)" of intellectuals cannot be (mis-)taken as a "source" or "influence." Nevertheless, in "How to Read," he says:

> I have been accused of wishing to provide a "portable substitute for the
> British Museum," which I would do, like a shot, were it possible. It isn't.
> . . . It struck me that the best history of painting in London was the Na-
> tional Gallery, and that the best history of literature, more particularly of
> poetry, would be a twelve-volume anthology in which each poem was cho-
> sen . . . because it contained an invention. (LE, 16–17)

In another context, Pound associates Gourmont's name with this "British
Museum project," when, in response to Sarah Perkins Cope, an under-
graduate who requested a reading list, he says: "My generation needed
Remy de Gourmont. Yeats used to say I was trying to supply a portable
substitute for the British Museum!!!" (L, 257). Critics have read such
comments about museums and anthologies as programmatic statements
of Pound's theory of direct presentation or reference to the text itself.
This is grossly to underestimate Pound's understanding of the politics of
museums and publishing and to disregard his careful attention to processes of selection and arrangement. An exemplary case of this concern
is his hope of replacing Palgrave's *Golden Treasury* with the *ABC of
Reading*'s "Exhibits," the section title alluding to both laboratory and
museum exhibits. This does not illustrate his belief in, say, the power of
poetry itself to civilize; rather, it proves his awareness of the rhetorical,
pedagogic, and political force of collections. Similarly, *The Tempio* "reads"
the Vatican or the Louvre, which, on the basis of Pound's commentary,
are shown to have as many strata, as much rhetoric, as *The Tempio*.

In this way, the various temples, museums, and anthologies scattered
throughout the texts of Spencer, Gourmont, and Pound expose the "rhetoric"—or textuality—of museums and System. Not only do museums
rely on complex interpretations of aesthetic value and appropriateness;
they are erected as monuments to a particular (Idea of) civilization or
government—that is, to what Pound came to call, borrowing from Plato
as well as Frobenius, *Paideuma,* a coinage to be fully explored in Chapter
Five. In other words, museums represent not only the art of the past but
also the self-reflections of their present donors, fundraisers, and the ap-
preciators of Art. They are, then, at least doubly rhetorical (or "rhetorical" in de Man's two senses, "figures of cognition" and "lies"): they are
synechdoches for Art and misrepresentations of the powers who erected
them. If he would not have approved of the terms of rhetorical criticism,
Pound was well aware of the stakes of the argument, and his own rhetorical figures affiliate him with Gourmont's more thorough examinations
of Spencer's method.

Pound does not comment on Gourmont's use of Spencer, nor is the
English naturalist on his list of favorite interpreters. Yet the names Gour-

mont and Spencer are linked in his text, in a quotation that seems to announce the project of dissociation—or deconstruction—as the analyses of the principal concepts and metaphors that underwrite both sociological and literary interpretations:

> La saine Sociologie *traite de l'évolution à travers les âges d'un groupe de métaphores, Famille, Patrie, État, Société, etc.* Ces mots sont de ceux que l'on dit collectifs et qui n'ont en soi aucune signification, l'histoire les a employés de tous temps, mais la Sociologie, par d'astucieuses définitions, précise leur néant tout en propogeant leur culte. . . . A la Famille, à la Patrie, à l'État, à la Société, on sacrifie des citoyens mâles et des citoyens femelles. (LE, 347; emphasis added)

Here Gourmont attributes to Spencer a rhetorical subtlety that has escaped all other readers and appears nowhere in the naturalist's most widely read sociology text, *Descriptive Sociology.* One does not know to which edition or translation Gourmont could have referred, but the version Pound would have known fits the description of a disorderly museum much more closely than that of a rigorous philological or rhetorical analysis. This eight-volume work, issued over as many years (1873–1881), consists of several three-by-three-and-a-half-foot taxonomical charts of the major accomplishments of English society under such headings as statecraft, family life, the arts and sciences. These are followed by pages of citations which presumably explain or analyze the data on the charts. The whole series of volumes, in so huge a format that it fits on no library shelf (not even on the coffee table, the usual place for pretty, unread volumes), was dedicated to Spencer's "American readers," and, if American readers and writers can be characterized by their "habitual quotations," as Pound suggested, this text and its intended audience are perfectly suited.

Spencer's section titles, including "Bias of Education," "Bias of Patriotism," and "Aesthetic Bias," indicate rhetorical analyses. His divisions and examples are indeed metaphors (metaphors of metaphor that mark the transitive nature of interpretations), but Spencer uses them as unproblematic categories into which he places his somewhat questionable data. His text is composed of citations from literary texts, government documents, and the interpretations of various modern experts. Spencer's sparse commentary suggests the format of the *Oxford English Dictionary,* which presumes to give first usages by way of quotations; it too relies on highly mediated texts.

Thus, if Frobenius's "kinematic map" represents an introduction of temporal, and rhetorical, disturbance into the stable taxonomies of classical history and anthropology, Spencer's sociology demolishes the logic of classification by the sheer weight of detail and the layers of translation and interpretation he forces taxonomy to bear. For example, by way of analyzing Elizabeth I's introduction of classical literature and the Italian Renaissance to the English court, Spencer merely quotes Hippolyte Taine's doctrine of the social benefits of art, which has Elizabeth as one of its examples: "'Elizabeth, half Gothic and half Italian, whose convenience, grandeur, and beauty announced already habits of society and the taste for pleasure. They [the courtiers] came to court and abandoned their old manners.'"[33] Thus, Elizabeth advances civilization by violating several historical definitions: she marks the epochal and categorical change from "Gothic" to "Renaissance" and the translation of ideas and culture from Italian to English. She becomes what Pound calls (borrowing, as we will see, from Jacob Burckhardt, that other nineteenth-century historian who can be argued to have influenced Nietzsche) and interpretative detail rather than a category, accounting for a radical cultural leap in history rather than for an evolutionary continuity and requiring interpretations rather than halting or authorizing a particular interpretation. This example might be thought to illustrate the Spencerian notions of individuation and the sudden disruptions aesthetic effects produce in society—or the most important aspect of compound evolution, differentiation. While they indicate progress, such transformations work against Spencer's own efforts to maintain orderly chronologies and spatially coherent taxonomies. Where does a figure like Elizabeth belong in the strict hierarchy and exclusive categories of taxonomical representation? And, of course, this is only one instance of the questionable evidence provided by quotations, translations, and unauthorized interpretations of *original* (but usually quoted) data. All these things belie any underlying unity, even one as flexible as Spencer's evolution, and frustrate any search for teleology.

Pound's reading texts often present similarly transformative persons or moments, and he tends to underscore the methodological problems they entail, though he sometimes christens such privileged cruxes as his own method(s): "luminous detail," "ideogrammic method," and "dissociation" all address such "facts." Instances of quotations which consist of layers of interpretations and translations standing unaccommodated in the Poundian text, most transposed from more systematic discourses, are legion—especially in those texts which promise to teach "How to Read" or offer a "Guide to Kulchur." An example from the early text "I Gather the Limbs of Osiris" is representative—as representative as anything can be of such heterogeneity. Here he acknowledges borrowing an economic principle as well as a historical crux from Jacob Burckhardt, the "inventor" of the Renaissance, which he defined, in sharp contrast to Pound, as the salvation of Classical values from the religious superstition, meaningless wars, and decadence of the Late Middle Ages: "When in Burckhardt we come upon a passage: 'In this year the Venetians refused to make war upon the Milanese because they held that any war between buyer and seller must prove profitable to neither,' we come upon a portent, the old order changes, one conception of war and of the State begins to decline. The Middle Ages imperceptibly give ground to the Renaissance" (SP, 22). If only coincidentally, Pound's reading of the German historian focuses on one of Spencer's main sociological categories, the state. But, more important than such overlappings, which are inevitable in the compendious texts of Pound and Spencer, is that Pound capitalizes on the discursive and categoric disruptions of the quotation (and of citation generally), drawing "economic" and "political" conclusions from a "historical" detail and using these to define aesthetic epochs. Perhaps not accidentally this particular transformative moment in the Late Middle Ages is contemporaneous with Malatesta's building of *The Tempio*. Such cross-discursive borrowing and troping is apparent in Pound's use of Gourmont's biologism as a storehouse of images for cultural and aesthetic principles. In later works Pound will turn to American cultural historians like Brooks Adams in order to explore the "economy" of transformations.

Pound's easy movement among discourses and genres characterizes his text from the start and becomes increasingly identified with his search for a unifying principle—or method—which might account for the inherently disseminative force of poetry. In Chapter Two, we saw his various borrowed methods and classifications put under extreme tension by his use of metaphors drawn from electromagnetics. Further, the various temples and museums that dot his text also suggest the way in which he allowed metaphor to play upon his reading method. Finally, *The Natural Philosophy of Love* offers what Pound claimed as not only a master principle which could unite all science and poetry but also (a figure for) "the very fluid of life itself." Thus, out of Gourmont's equation of seminal and cerebral fluids, intellection and instinct, he fashioned an account of the poetic medium and textual transmission. He offers, then, as the unifying principle underlying all the privileged details and individual geniuses of literature, a biological metaphor which stresses the dissemination of "images," not the orderly progress or evolution of "tradition"—the pun on *semen* and *seme,* pointed out by a recent critic, is appropriate to Pound's notion of a "spermatic economy," which might be said to be in a state of "run away inflation."[34] In other words, Pound again offers disruption and excess by way of system: poetic figures as definitions of poetry, a notion of metaphor which, as we shall see, works *more metaphorico*.

Moreover, he claims that his borrowed "hypothesis" ("Il y aurait peut-
être une certain corrélation entre la copulation complète et profonde et le
dévelopment cérébral") is more comprehensive and more "scientific"—
in his peculiar sense of a science, which is, like "the ideogrammic method,"
grounded in "natural metaphor"—than the reigning physiology, psychology, and epistemology of his time: "This hypothesis would perhaps
explain a certain number of as yet uncorrelated phenomena both psychological and physiological. It would explain the enormous content of the
brain as a maker or presenter of images" (NPL, 206). In a sort of circular
evolution (re-volution, re-turn or trope) which escapes both entropy and
monolinear progress, Pound's method departs from and returns to poetry,
and particularly to erotic lyrics, and "poetic metaphysics," both associated with Gourmont's own reading program:

> Perhaps the clue is in Propertius after all:
> > *Igenium nobis ipsa puella fecit.*
> There is the whole of the XIIth century love cult, and Dante's metaphysics
> a little to one side, and Gourmont's Latin Mystique; and for image-making
> both Fenollosa on "The Chinese Written Character," and the paragraphs
> in "Le Problème du Style." At any rate *the quarrel between cerebralist
> and viveur and ignorantist ends, if the brain is thus conceived not as a
> separate and desiccated organ, but as the very fluid of life itself.* (NPL,
> 219; emphasis added)

Like Spencer, Pound sought a sort of "Synthetic Philosophy," but one
consistently opposed to philosophic system. This de(con)structive war of
literature against philosophy—even of Nietzsche against Darwinism—is,
as Pound suggests, indebted to Gourmont, on at least three scores: *Le
Latin mystique,* "Le Problème du style," and, most importantly, *The Natural Philosophy of Love.* In the course of the "Translator's Postscript,"
he silently adopts Gourmont's transvaluations (deconstructions?) of several key terms in order to reassess the values of "instinct," "sexuality,"
"nature," and "creativity." These revisions underwrite Pound's *notion*
(the very antithesis of Eliot's *Idea* of tradition) of the revolutionary and
utilitarian aspects of poetic "image making," which is but one manifestation of the prodigious thought of the "seminal brain":

> Creative thought has manifested itself in images, in music, which is to
> sound what the concrete image is to sight. And the thought of genius, even
> of the mathematical genius, the mathematical prodigy, is really the same
> sort of thing, it is a sudden out-spurt of mind which takes the form de-
> manded by the problem; which creates the answer, and baffles the man
> counting on the abacus. (NPL, 212)

Like Spencer, Pound is all over the shop—or museum or university curriculum—and engaged by Darwin's model, in ways we have already considered. Further, his critiques of the transmission of texts and tradition by means of the "ejaculation" of thought, his "over*coming*," as it were, of the moral and genealogical prejudices of literary history, must be understood in the context of Gourmont's adoption of Nietzsche's transvaluation. This conjunction of Nietzscheanism and anti-Darwinism has been noted by Gourmont's editors and critics. Burton Rascoe, for example, suggests a link between transvaluation and Gourmont's reversal of the Darwinian teleology, which places the Victorian Christian at the summit of evolution.[35] Richard Aldington noted Gourmont's esteem for Spencer's scientism and his attraction to the flexibility of transvaluation. Yet such "associations" remain to be explored with regard to Pound's "reading" and Nietzsche's transvaluation, especially because the latter term is too readily accepted as commonplace and stable—that is, as a philosophical concept rather than a rhetorical strategy. Therefore, we should now pick up the thread of transvaluation as it runs through Nietzsche's text and, however unaccountably, through Gourmont's to Pound's.[36]

III

"Transvaluation of all values" is no simple formula; indeed, it names Nietzsche's whole interrogation of the grounding concepts of science and metaphysics. Like Will to Power, which can be thought of as a particularly militant phase of Nietzsche's deconstructive reading, it strategically disturbs established beliefs and Truth. It is not, however, nihilistic; rather, it affirms interpretation by abolishing—but not simply reversing—the "mastery" of philosophical or religious over aesthetic and sensuous "values." It would seem that Gourmont knew this sense of Will to Power but perhaps did not apply it with the full rigor of more recent French Nietzscheans. For example, Maurice Blanchot makes a useful distinction between transvaluation and that *nihilism* (or even modernism) which would simply replace old with new values: "Transvaluation does not give us a new scale of values on the basis of negating every absolute value; it makes us attain an order for which the notion of value ceases to apply."[37] Transvaluation equates "natural" and "cultural" phenomena, in order to show that all of these "things" are interpretations: *"There are no moral phenomena, there is only a moral interpretation of these phenomena"* (WP, 149). This is to say that scientists, philosophers, and artists are engaged in maintaining a nearly invisible system of moral values dictated by religion and the reigning behavioral norms: "Fundamental insight: Kant as well as Hegel and Schopenhauer—the skeptical-epochistic attitude as well as the pessimistic—have a *moral* origin" (WP, 221). It is Nietzsche's— "the new philosopher's"—special mission to uncover, or anatomize, the metaphysical assumptions which enforce Truth.

The best definition of transvaluation appears in that crucial section of *Ecce Homo,* "Why I Am So Clever," in a passage especially difficult for English translators. In fact, Walter Kaufmann has rendered the key term, *Umvertung der Werte,* "revaluation of all values," suggesting negation or simple reversal. Yet it should be kept in mind that *um,* the preposition meaning "about," "round," and "round about," has the sense of movement or troping (as in *Umwändlung,* or "transformation"). Nietzsche's definition, which exemplifies the *working*—the praxis, not the theory— of transvaluation, is not a revaluation of the philosopher's task but a rhetorical analysis of "philosophical disinterestedness," which has always been opposed to such things as (animal) instinct and the falsifications of art:

> For the task of a *revaluation of all values* [emphasis in text] more capacities may have been needed than have ever dwelt together in a single individual—above all, even contrary capacities that had to be kept from disturbing, destroying one another. An order of rank among these capacities; *distance;* the art of separating without setting against one another; to mix nothing, to "reconcile" nothing; *a tremendous variety that is nevertheless the opposite of chaos—this was the precondition, the long, secret work and artistry of my instinct.* (EH, 254; emphasis added)

As befits a work entitled *Ecce Homo,* Nietzsche here begins autobiographically, but his nearly biological ("capacities," "instincts") analysis undoes the notion of a unified "subject," let alone an authorial self. For Nietzsche, the philosopher is the highest man, not because of his proximity to truth or his resolution of all conflicting interpretations, but because he accepts and lives as many contradictions as possible. Just so, this very definition also addresses the "new science" which opposes intellection to instinct, and banishes art from philosophy and science. This is to say that Nietzsche's oxymoron "artistry of my instinct" (*Kunsterschaft mines Instinckts*) transvalues the Darwinian notion of evolution away from "the tyranny of instinct" toward intellection and culture.

While Gourmont cites neither *Ecce Homo* nor *The Will to Power*'s even more explicit anti-Darwinism, *The Natural Philosophy of Love* adopts a similar transvaluative strategy in order to make the point about Darwin's too great love of System: "Darwin arrived, inaugurated a useful system, but his views were too systematized, his aim too explanatory and his scale of creatures with man at the summit, as the culmination of universal effort, is of a too theologic simplicity" (NPL, 12). Like Nietzsche, Gourmont wants to introduce contradiction into systematic thought, or poetry into philosophy, in order to reverse the essentializing, and antisensual, tendencies of modern science. To this end, he protests the privilege afforded the intellect at the expense of instinct, in a passage that might be thought to echo Nietzsche's assaults on the false philosophers who, repressing their own desires, master the creative force of art: "Every organized animal has a master: its nervous system. . . . As soon as this new matter appears, it reigns despotically . . . and lifting itself into the royal rôle" (NPL, 204).

This centralized power, as it were, is redistributed in Pound's notion of an anarchical—or at least dispersed and multicentered—nervous-spermatic system: "The brain itself is, in origin and development, only a sort of great clot of genital fluid held in suspense and in other species, held in several clots over the scattered chief nerve centres" (NPL, 206). Such a multiplication of centers—analogously, of literary sources, methods, and interpretations—reflects Pound's desire to supplement or reinvent tradition, not to abolish it.

At several points in his two hundred pages of nearly pornographic descriptions of insect copulation, Gourmont makes programmatic statements—or, better, statements critical of scientific program—which Pound borrows for his own long-evolving (de-)definition of culture. Not all of these are confined to the "Translator's Postscript." For instance, Gourmont's transvaluation of instinct, if not Nietzsche's "artistry of instinct," appears at a crucial point in *Guide to Kulchur* (written in 1938, sixteen years, that is, after his translation of *The Natural Philosophy of Love*). Here Pound equates Gourmont's assaults upon the exclusive privilege of intellect with Frobenius's notion of *Kultur* as a residuum: "'Culture: what is left after a man has forgotten all he set out to learn?' Cf. Gourmont's 'instinct' as a result of countless acts of intellection, something after and not before reason" (GK, 195).[38] While Pound does not cite a specific passage in this regard, one might suppose that, as he pondered Gourmont's every word for its translation into American, he might have paused over this passage, which accuses science and common sense of the sort of rhetorical obfuscation he everywhere descries:

> The question of instinct is perhaps the most nerve-racking there is. . . .
> In the vulgar contrast one overhears the considerable naiveté that animals
> have instinct and man, intelligence. *This error, pure rhetoric, has pre-
> vented,* up to the present, *not the answer* to the question which still seems
> a long way off, *but the scientific exposure of the question itself.* It includes

but two formulae: *Either instinct is a fructification of intelligence; or intelligence is an augmentation of instinct.* (NPL, 184–85; emphasis added)

As we have seen, Pound poses the question of renewing artistic creativity and resolves the conflict of instinct and intelligence by using the figure of a "fructifying intelligence," specifically in the metaphor of a "spermatazoic brain." In this way, he engages Gourmont's Nietzschean critique of scientific method in the most profound way; that is, indirectly, metaphorically. In order to grasp the methodological implications of such a reading, for which Pound provides no commentary, we again have recourse to Nietzsche.

Nietzsche proposes analyzing obfuscating moral concepts by attending to life's "small things"—to Nietzsche's *"kleines Dinge,"* not *Ding-an-sich.* Gourmont might be thought to take Nietzsche a bit too literally when he focuses on minutiae of insect behavior to expose the prejudices of science, but his analysis uses these strategies:

> Small things—nutrition, place, climate, recreation, the whole causistry of selfishness—are inconceivably more important than everything one has taken to be important so far. Precisely here one must begin to *relearn.* What mankind has so far considered seriously have not even been realities but mere imaginings. . . . "God," "soul," "virtue," "sin," "beyond," "truth," "eternal life."—But the greatness of human nature, its "divinity," was sought in them.—All the problems of politics, of social organization, and of education have been falsified through and through because one mistook the most harmful men for the great men—because one learned to despise "little" things, which means the basic concerns of life itself. (EH, 256)

While Nietzsche takes aim at far-reaching political, social, and pedagogic issues, Gourmont ostensibly limits his analysis to the insect world. Nevertheless, by reading anthropomorphized natural specimens alongside the text of nineteenth-century natural philosophy, he extends Nietzsche's critique of philosophy's "moral falsifications" to discourses as disparate as literature and entomology. And, like Nietzsche, he does this by the careful rhetorical analysis of the names science bestows on natural phenomena; that is, he exposes science's "facts" as complex interpretations—if not simple "fictions." Rather than rejecting science, he causes poetry and rhetoric to intervene in scientific method. Nietzsche says, "It is not the victory of science that distinguishes our nineteenth century, but the victory of scientific method over science" (WP, 261).[39] Though less rigorously than his precursor, who had refused to make the categories of Poetry and Science simple binary opposites, Gourmont enters this pitched battle on the side of poetry. His undoing of English naturalism by ref-

erence to "French thought" is a case in point—and, in passing, one might
note that Pound not only praises this French habit of mind, but he similarly finds in literature a scientific precision absent from science itself:

> M de Sade liked to imagine ruttings where blood and sperm flowed si-
> multaneously; mere kindergarten manual (Berquinade) if one contemplates,
> not without bewilderment, the habits of an ingenious orthopter, *the praying
> mantis, the insect which prays to God, la prego-Diou as the Provençals
> call her, the prophetess as the Greek said! Baudelaire's verses* ridiculing
> those who wish to
>   "aux choses de l'amour meler l'honnêteté"
>   Mix seemliness into affairs of Love
> *have a value not only moral but scientific.* (NPL, 77; emphasis added)

Here Gourmont's point is hardly the Arnoldian commonplace that poetry represents man's "best," which is to say, most "moral" self, but that
poems can be used as critical tools against the moralizings of literary
critics as well as those of scientists. By thus invoking the "unnatural" or
pan-natural philosophy of Sade and the decadent poetry of Baudelaire,
he mocks the Darwinian attribution of piety to the *praying* mantis, a modern personification that is only the latest instance in a whole etymology
of sentimental names which fictionalize insect onanism, sodomy, can-
nibalism, and their other "natural aberrations" that outstrip Sade's fan-
tasies. It would seem that Gourmont also works a subtle parody of that
favorite trope of philologists, which is turning names into Truth by showing the frequency and antiquity of a particular usage.[40]
Throughout his text Gourmont similarly calls upon his favorite literary
figures to mock scientific prudery. Rather than simply correcting the misprisions of science, or replacing these with direct observation, he criticizes the scientific tendency to use hackneyed and imprecise metaphors.
He wants science not to stop writing fiction or using poetic language, but
to start writing more accurate poetry. Thus Sade, that French writer marginal to both literature and science, appears in this ironic reading of J.
H. Fabre's entomology:

> The female philanthe will attack the nectar-loaded bee with her great dart,
> stab him and pump out his crop. One may see the ferocious small animal
> knead the dead bee for half an hour, squeeze him like a lemon; drink him
> out like a gourd. Charming and candid habits of these winged topazes whir-
> ring among the flowers! *Fabre has excused this sadique gourmandizing:
> the philanthe kills bees in order to feed her larvae,* who have, however,
> so great a repugnance for honey that *they die upon contact with it; it is
> therefore out of sheer maternal devotion that she intoxicates herself with
> this poison!* (NPL, 37–38; emphasis added)

In a somewhat more serious vein, Gourmont deploys his rhetorical strategy against the most privileged judgments of science and theology, the "natural" and "normal." Arguing that libertinism is a symptom of higher intelligence, that promiscuity raises man beyond the mechanical demands of reproduction (both sexual and technological), he shows that morality, an arbitrary and repressive system of signs, has no foundation, and language itself does not originate in (the things of) nature: "It is very difficult, especially when dealing with man, to distinguish between normal and abnormal. What is the normal; what the natural? Nature ignores this adjective" (NPL, 79). The following passage summarizes Gourmont's analysis of human instinct and intellect, which he sees converging in the imaginative artistry of ("unnatural") human sexual pratices: "One must not be gulled by the scholastic distinction between instinct and intelligence; man is as full of instincts as the insect most visibly instinctive; he obeys them by methods more diverse" (NPL, 15). Pound translates the notion of creative sexuality into the aesthetic realm when, for example, he suggests that all the instincts, capacities, even the aberrations of the animal kingdom are available to man—specifically to the poet, who most readily creates forms and images:

> Man is the sum of the animals, the sum of their instincts. . . . Given, first a few, then as we get to our own condition, a mass of these spermatozoic particles withheld. . . . Each of these particles is, we need not say, conscious of form, but has by all counts a capacity for formal expression: is not thought precisely a form comparing and form combining? (NPL, 211)

Both Gourmont and Pound would seem thus to argue that Darwin and the humanists correctly place man at the end of the evolutionary process, but for all the wrong reasons. Man—and especially the artist—holds this position by virtue of transvaluing and therefore reintroducing sexual force and other instincts, not by evolving beyond them. Man is, by this account, a compendium of desires and the nearly infinite ways of satisfying them: "Man is the sum of the animals, the sum of their instincts" (NPL, 211), and even an excess of any sum. Art, the agent of change and progress, unsettles bourgeois society and morality; thus Gourmont's biological-aesthetic notion of individuation and genius: "Collective civilization has diminished the individual genius" (NPL, 193). This modern war of the individual artist against convention has a long and not easily defined history, which, as we have seen, includes Nietzsche, Spencer, and the Symbolists.

For his part, Pound announces this theme by invoking a whole poetic-scientific tradition of phallocentrism when he equates his poetic and, it would seem, his pedagogic (as well as journalistic) efforts on behalf of new ideas with "the male feeling in copulation":

> There are traces of it in the symbolism of phallic religions, man really the phallus or spermatozoide charging, head-on, the female chaos. Integration of the male in the male organ. Even oneself has felt it, driving any new idea into the great passive vulva of London, a sensation analogous to the male feeling in copulation. (NPL, 207)[41]

Sexism aside (and Pound does graciously allow "a cognate role to the ovule"), the point here is that modern society, and scientific thought particularly, requires an infusion of new ideas—new poetic forms and images, as he makes clear, not Idea. Notwithstanding the self-satisfied characterization of his own role, he here abbreviates his plan of revitalizing a worn-out tradition in a way which, as we will see, hardly fits the old genealogical (also phallocentric) order of philosophy and literary history. He suggests, then, that every epistemological moment is also a sexual one; but, more to the point, both are textual or imagistic—meta*morph*ic or, more simply, metaphoric:

> I offer an idea rather than an argument, yet if we consider that *the power of the spermatozoide is precisely the power of exteriorizing a form;* and if we consider the lack of any other known sbustance in nature capable of growing into brain. . . . Given, that is, two great seas of this fluid [in the two lobes of the brain], mutually magnetized, the wonder is, or at least the first wonder is, that human thought is so inactive. (NPL, 206–7; emphasis added)

It remains to be seen how Pound turns his notions of sexual instinct, poetic genius, and imagistic—as opposed to abstract—thought into a "theory" of textual transmission, if not a deconstructive reading of literary history. We should turn, at this point, to his "Translator's Postscript" and to its extended scientific-physiological metaphors, which he used to intervene in scientific method and the tradition.

IV

Pound's efforts to reverse the exhaustion and chaotic dispersal of thought by introducing new images are directed at the triple threat to modern life, as he saw it: the nihilism of nineteenth-century science, the abandonment of poetry for the mechanical, and the preference for abstract thought and language in general. At one point, he confesses, "I appear to have thrown out bits of my note somewhat at random" (NPL, 216). While one is accustomed to such apparent randomness in even his programmatic essays, Pound's "Translator's Postscript" nevertheless comprehends the major concerns that underwrite Gourmont's deconstructive reading program. Moreover, he extends these interests to the province of his own rhetorical—tropological—notion of Imagism and to the affirmation, recurrent throughout the whole of his reading program, of culture, and especially the poetic tradition, as a dissemination of meaning. At no point in this essay does he simply reject scientism for an originary or primordial poetry or philosophy for a more authentic and adequate expression. Instead, he *disrupts*—one can say "transvalues" even "deconstructs"—scientific theories and terminology from within, borrowing and bending them to his own literary critical ends. Therefore, he also disturbs the jingoist dreams of nativist or American poetry and poetics which, notwithstanding repeated denials of formalism, draw the line at such methodological promiscuity.

For example, he toys with evolution by suggesting that the process has yet to fulfill itself in the development of a new (poetic?) organ or faculty, which would issue from the spermatic excesses of genius: "Not considering the process ended; taking the individual genius as the man in whom the new access, the new superfluity of spermatozoic pressure (quantative and qualitative) up-shoots into the brain, alluvial Nile-flood, bringing new crops, new invention" (NPL, 217–18). This new invention could be either scientific or a rhetorical *inventio* or a new bodily part or organ. The same "form-creating" fluid or faculty (the "spermatozoide") that produces new images or "thought-forms," he argues, effects the sudden appearance of new organs that advance biological species: "the langouste's long feelers, all sorts of extravagances in nature may be taken as the result of a single gush of thought. A single out-push of a demand, made by a spermatic sea of sufficient energy to cast such a form" (NPL, 213–14).

The appearance of any new organ, then, would be an act of the spermatic or form-creating brain. By contrast to natural selection, such physiological advances are the results of conscious acts, like reading and writing: "Remains precisely the question: man feeling this protean capacity to grow a new organ: what organ? Or new faculty; what faculty?" (NPL, 215). He then provides a cryptic answer, which only suggests the poetic or form-creating faculty: "*The new faculty? Without the ostentation of an organ. Will?*" (NPL, 216; emphasis added). In lieu of a definition of "Will," Pound generates a list of things not worth willing—or things which stand in the way of "Will" understood as creative thought: "The hypnotist has shown the vanity and Blake the inutility of willing trifles, and black magic its futility. . . . Take it that what man wants is a capacity for *clearer understanding,* or for *physical refreshment and vigour, are not these precisely the faculties he is forever hammering at, perhaps stupidly?"* (NPL, 216). Pound does not seek the moral or hygienic improvement of man either in poetry or physiology. Instead, he might be thought to parody Darwinian notions of progress and simple utility, though his proposed new faculty remains enigmatic.

It is tempting to suppose that Pound alludes to Nietzsche's Will, which is tied up with poetry, interpretation, and the force of these. As Joseph Riddel has convincingly argued, his notion of form-creating thought nearly repeats section 805 of *Will to Power.*[42] But, quite independent from Nietzsche and the other possible sources we have considered, Pound indicates the tropological character of this faculty—call it will or writing or reading. It is the ability, at the exhausted end of the literary tradition and the thermodynamic cycle, to return to poetry and thus save man from the mechanical:

> In his growing subservience to, and adoration of, and entanglement in machines, in utility, man rounds the circle almost into insect life, the absence of flesh; and may have need even of horned gods to save him, or at least a form of thought which permits them. (NPL, 209)

Pound implies the superiority of mythopoeic (metamorphic and highly metaphoric) to scientific thought, suggesting that industrial man, surely one of Darwin's exemplary specimens, reverses the hierarchy he ostensibly surmounts, returning, through "absence of flesh" and sexuality, to the lowest life forms. This is not to say, however, that Pound calls for a return to a simpler, pastoral life. On the contrary, the new faculty he proposes would seem a most unnatural mixture of the senses and certain other knacks for reading and writing otherwise.

For example, his list of modern prodigies that might advance the evolutionary process and improve culture is a motley assemblage of literary allusions and at least one whimsical "reading scene":

> You have every exploited *"hyper-aesthesia," i.e., every new form of genius, from the faculty of hearing four parts in a fugue perfectly, to the ear for money (vide Henry James in "The Ivory Tower"* the passages on Mr. Gaw) . . . and you have *Mr. W.,* a wood-broker in London, who suddenly at 3 A.M. *visualizes the whole of his letter-file,* three hundred folios; he sees and reads particularly the letter at folder 171, but he sees *simultaneously the entire contents of the file, the whole thing about the size of two lumps of domino sugar* laid flat side to flat side. (NPL, 215; emphasis added)

It is noteworthy that Pound has metaphorically expanded the definition of "hyper-aesthesia," which means the pathological sensitivity of the skin and, by (his) extension, all the senses, as a new form of genius. This usage is, in fact, consistent with his equation of physiological and aesthetic faculties in the figure of the spermatic brain. Moreover, this passage seems to call for the cross-discursive interests he found in the great literature of the past: if in James he seems to find an ear for money, in the *Divine Comedy* he reads a "single extended metaphor for life." Thus, this strange synesthesia, of an ear both for music and for money and Mr. W.'s vision of thousands of facts metonymically referenced in one, is involved, however indirectly and figuratively, in his larger reading concerns. Surely, Mr. W.'s form of genius cannot be taken entirely seriously, yet Pound elsewhere praises such concision and retention, if he would put this prodigious loose-leafing or filing system to better use than the wood-broker. Perhaps it is too far-fetched to suppose that this file adumbrates Pound's characterization of his own reading text, as in, for example, "How to Read," where he says: "We apply a loose-leaf system to book-keeping so as to have the live items separated from the dead ones. . . . and we could, presumably, apply to the study of literature a little of the common sense that we currently apply to physics or to biology" (LE, 18–19).

A little further on in the "Postscript," he again raises the question of the new faculty in connection with reading. If his examples seem serious, the precise character and value of reading remain unclear. His description of the passive (which is to say feminine) yet evaluative and discriminating nature of reading suggests his own ambivalent relationship to tradition and innovation:

> The dead laborious compilation and comparison of other men's dead images, all this is mere labour, not the spermatozoic act of the brain.
>
> Woman, the conservator, the inheritor of past gestures, clever, practical, as Gourmont says, not inventive, always the best disciple of any inventor, has always been the enemy of the dead or laborious form of compilation, abstraction. (NPL, 217)

We cannot do justice to the apparent sexual ambiguity of the reading act, but Pound does not maintain a clear opposition between passive and active, good and bad, female and male reading. Here he would seem to suggest, at the very least, that not all "compilation" and "comparison," even of inherited images, need be laborious or unproductive.

Notwithstanding his full treatment of spermatic thought and the "ejaculation" of poetic images, Pound does not finally resolve the issue of the relative values of reading and writing, of continuity and revolution. This same indecision troubles his categories of "master" and "inventor" in the reading textbooks, thereby confirming his notion of "creative thought" as inherently disseminative—which is to say interpretive and subject to infinite, open interpretation. Indeed, he aims in the "Postscript" to show that science and philosophy, which insist on the clear delineation of sexual and textual differences (on such fundamental "facts" as sexual dimorphism and the generic exclusivity of the various arts and sciences) cannot withstand the force of creative genius. Thus, on Gourmont's authority, he directs his description of the advent of new ideas, as we noted above, against the limiting principle of Reason; and, in the same gesture, he frees his physiology from moral and discursive constraints:

> Not considering the process ended; taking the individual genius as the man in whom the new access, the new superfluity of spermatozoic pressure . . . up-shoots into the brain, alluvial Nile-flood, bringing new crops, new invention. And as Gourmont says, there is only reasoning where there is initial error, i.e., weakness of the spurt, wandering search. . . . This is a question of physiology, it is not a question of morals and sociology. (NPL, 217–18)

In this way, Pound challenges the inherent nihilism of the "new sciences" of geology and biology, which subscribed to the notion that man and the physical universe had nearly arrived at the end of the evolutionary and thermodynamic lines—an end ironically near an apex—and were beginning their abrupt decline. He tries to reverse this entropy, not only by proposing new biological developments but by holding out the possibility that "enough energy" remains from "the thermodynamic cyclone" to allow "vegetable" or "form-creating thought" through the "cracks left by the earth's cooling process" (NPL, 212, 217). He does not thereby reject the reigning scientific assumptions; he merely takes them a step further, as it were, and into an argument for the overcoming of science. His richly figurative account of the redemption of science by poetry becomes a rhetorical critique of the abstract and imprecise (*entropic* tropeless, troped-out or wornout) language of science: for example, he distinguishes imaginative—even Imagistic—thought from ratiocination which he characterizes, playing on the double sense of "ratio," as "the least common denominator of a multiple of images." I quote the passage containing this comparison at some length, because it typifies the variety of tropes he deploys against abstraction—including puns, clichés, comically mixed metaphors or catachreses, and folk etymologies, or all those things generally abhorred by serious critics:

> We have the *hair-thinning "abstract thought"* and we have the *concrete thought of women, of artists, of musicians, the mockedly "long-haired,"*

who have made everything in the world. We have the form-making and
the form-destroying "thought," only the first of which is really satisfactory.
I don't wish to be invidious, *it is perfectly possible to consider the "ab-*
stract" thought, reason, etc., *as the comparison, regimentation, and least*
common denominator of a multiple of images, but in the end each of the
images is a little spoiled thereby, no one of them is the Apollo, and the
makers of this kind of thought have been called dry-as-dust since *the be-*
ginning of history. (NPL, 211; emphasis added)

Basing arguments on such questionable "evidence" as the physiognom-
ic clichés of "long-haired" and "dry-as-dust" is, in fact, typical of Gourmont's biology, which marshalls literary allusion and "the common wisdom" against the self-important seriousness of modern philosophy.
Similarly, Pound carries on his argument against science in rhetorical
terms by juxtaposing his privileged images with the abstract language of
science, which has failed to grasp the underlying Truth of poetry—failed,
for example, to recognize the value of even the merest traces of a whole
complex of symbols and gods from the phallic religions. He also claims
that biology and poetry have spermatic thought as their common source,
that images from the "spermatic brain," even the clichés or dead metaphors of science, are preferable to abstract or philosophical thought.

He repeats one of the oldest metaphysical arguments: that language,
and hence all the discourses of men, fell from the unity of an originary
Poetry into the various arts and sciences that only approximate Truth.[43]
As we have seen, Derrida, in the "*Retrait* of Metaphor," links philosophy's nostalgia for, and repression of, figurative (or "poetic") language
with the most powerful of Western figures, the sun, which is also the
"helio-trope," or Apollo, the Greek sun god. Perhaps not entirely by accident, Pound invokes Apollo against the abstract language of philosophy
and science. It would seem, then, that his talk of "horned gods," "Apollo,"
and "the beginning of history" affiliates him with the Western (onto-theo-
logic) philosophy he criticizes. He seems intent on recuperating an originary or natural poetry—even an organic one, grounded in the very blood
and sperm of man and thus *before* language. Nevertheless, this return to
the primordial, to Poetry, or Truth, is everywhere frustrated by the proliferation of images and, what is more important, by his insistence on
this excess. Thus, while we might have missed the allusion out of context, Apollo is the proper name of one of Gourmont's favorite butterflies
as well as the Roman name of the Greek god.

Moreover, Apollo is not the only god Pound invokes; in fact, his strange
mythology involves a certain excess: an intervention of the poetic in science and philosophy, which sometimes goes under the name of Diony-

sus—even Proteus. In this way, Pound suggests a poetics in which the
Apollonian and Dionysian are inextricably interwoven at the (non-)origin
of poetry and/or in the textures of a "form-creating thought." He introduces Dionysus or Bacchus, the horned god, as a resistance to entropic
modern thought, which "may have need even of horned gods . . . or at
least a form of thought which permits them" (NPL, 209). Such "thought"
would necessarily involve the two aspects of intellection which Pound,
following Nietzsche almost to the letter, calls both "form-creating" and
"form-destroying." This ambiguous force, at once abolishing any constraints placed on poetry and creating or casting out more poetic images
than any one text can master, recurs several times in *The Cantos* and,
perhaps not altogether fortuitously, under the guise of the horned god—
or the transformational Zagreus, Dionysus, Digones, Bacchus. For instance, in Canto II, Pound relaunches his (Apollonian?) epic by switching
from Homer's ships to Ovid's bark, which transports an unthinkably
metamorphic cargo of quotations, translations, and interpretations super-
added to the transformational stories of Golding's *Metamorphoses*. His
drunken boat, (reminiscent of Rimbaud's?) carries the god of wine away
from his "home" in Naxos, but this delay or detour enables the irrepressible force of sexuality and creativity to generate the uncontainable poetic
images that comprise Pound's poem and reflect—or refract—his polemical tradition. In spite of the sailor's efforts to enchain him, Dionysus
creates a new text out of an old figuration:

> And where was gunwale, there now was vine-trunk
> And tenthril where cordage had been,
>   grape-leaves on the rowlocks,
> Heavy vine on the oarshafts,
> And, out of nothing, a breathing,
>   hot breath at my ankles,
> Beasts like shadows in glass,
>   a furred tail upon nothingness.
>
>   (II, 8)

By means of a palimpsest of texts—even of "shadows in glass," but
hardly with the alleged self-reflective transparency of a "mirror held up
to nature"—Dionysus marks one of *The Cantos'* epistemological yet textual moments of "form-creating" and "form-destroying" thought.

In this way, Gourmont's text and Pound's "Translator's Postscript"
work, in conjunction with the other reading notes and textbooks, not only
to produce a reservoir of definitions or, better, dissociations that make
poetic metaphors out of biological metaphors, but tend also to produce
translative effects that resonate in his later poetry and critical prose. This

is hardly a "closed economy" or systematic theory of either the Image or the long poem, not a method either for reading or writing. Rather, it can be thought to illustrate the sudden "out-spurt" and "alluvial Nile-flood" of a spermatic economy, something like that which Derrida has come to call "disseminative" translation.[44] As might be expected from definitions which partake of the force they describe, examples of Pound's cross-referencing proliferate (as in Derrida's rhetorical analyses which generate tropes *more metaphorico*), but one more example will suffice to show the coincidence of Dionysian thought and Pound's efforts to complete science and mathematics by a reinfusion and reinscription of poetry. The definition of "creative thought" we looked at earlier equates mathematical with poetic "genius," but this genius exceeds both discourses: "Creative thought has manifested itself in images. . . . The thought of genius, even of the mathematical genius . . . is really the same sort of thing, it is a sudden out-spurt of mind . . ʼ. which *creates the answer, and baffles the man counting on the abacus*" (NPL, 212; emphasis added). Years later, in a passage from *Thrones* which catalogues the virtues of Medieval numismatics, Pound recalls this disruptive calculation; again, the god of wine plays with the imagination (the image making faculty) or, better, disrupts grammar with the poetic:

> *a gold Bacchus on your abacus,*
> Henry Third's second massacre, wheat 12 pence a quarter
> that 6 4/5th pund of head be a farden
> Act 51, Henry Three. If a penny of land be a perch
>   *that is grammar*
> *nummulary moving toward prosody.*
>   (XCVII:671; emphasis added)

However freely or illogically, such associations illustrate the workings of Pound's "reading machine" which operates according to the imagistic economy it defines. One cannot claim that Pound raises the spermatic model of textual production and transmission to the level of *theory* (after all, such a theory could be nothing more than the impossible hybrid of a "meta-metaphorics"), yet it is possible to see his "form-destroying" rhetoric as a kind of Nietzschean force, undoing the God of grammar with the gods of transformation. Pound's casual Dionysus repeats Nietzsche's distinction between the Apollonian and Dionysian, that is, between the god who creates forms and the god who unleashes the metamorphic force out of which forms are created. Not unlike Pound's, Nietzsche's formulation inscribes the Apollonian in the Dionysian (or is it the other way around?):

Apollonian intoxication alerts above all the eye so that it acquires power of vision. The painter, the sculptor, the epic poet are visionaries *par excellence. In the Dionysian state,* on the other hand, *the entire emotional system is alerted* and intensified; so that *it discharges all its power of representation, imagination, transfiguration, transmutation, every kind of mimicry and play-acting, conjointly. The essential thing remains the facility of metamorphosis.* . . . It is impossible for the Dionysian man not to understand any system of whatever kind, he ignores no signal from the emotions, he possesses to the highest degree *the instinct for understanding and divining, just as he possesses the art of communication to the highest degree.* (TI, 73; emphasis added)[45]

Nietzsche's Dionysian Man appears the very *summa* of instinct and understanding, inarticulate emotion and critical acumen, that Pound described as the cultural instinct which culminates art and intellection. And Nietzsche's Dionysian "discharges"—in an "out-spurt" of signs, of *semes,* if not semen—the figurative power of language in a virtual anthology of tropes. His rhetoric, which he ironically calls "metamorphosis," or that which cannot be present-to-itself, an "essential thing," or that which must be, dismantles all notions of re-presentation: "imagination, transfiguration, transmutation, every kind of mimicry."

If Nietzsche's account seems more explicitly aesthetic—or, more exactly, a rhetorical critique of (Wagner's) "philosophical art" and systematic (that is, Hegel's) *aesthetics*—than Pound's physiology of style, the "Postscript" nevertheless addresses the *same* (yet differential) dissemination of images:

Thought is a "chemical process" in relation to the organ, the brain; *creative thought is an act like fecundation, like the male cast of the human seed,* but given that cast, that ejaculation, *I am perfectly willing to grant that the thought once born, separated, in regard to itself,* not in relation to the brain that begat it, *does lead an independent life* much like a member of the vegetable kingdom, *blowing seeds, ideas from the paradisal garden at the summit of Dante's Mount Purgatory, capable of lodging and sprouting where they fall.* And Gourmont has the phrase "fecundating a generation of bodies as genius fecundates a generation of minds." (NPL, 210–11; emphasis added)

Here Pound, while dangerously close to repeating the old phallocentrism or paternalism he elsewhere seems quite comfortable with, transvalues the genetic model of tradition, which depends on the orderly transfer of texts from Classical writer to modern reader, as from father to son. With the same gesture, he unsettles the conventional notion of the organic or "vegetable" which inscribes a self-contained network of ideas. For Pound, the highest Art is unpredictable and uncontainable: the element of continuity, whether "spermatic thought" or the "luminous details" of Dante's poetry, is also the vehicle of change—and the medium of exchange. But this is not to say that Pound's notions of "reading" and *Kulchur,* however de(con)structive they might be of Eliot's tradition, are either immune to change or fully open to it.

CHAPTER FIVE

Kulchural Graphics
Paideuma and the Guidelines to Poetry

"Gorgias debunks the logical process"—
Headline on Athenian billboard.[1]

What I should do would be a long essay, criticism of Greek and Latin cultural heritage *confronted* by post-Renaissance knowledge of subjects not familiar to Pico della Mirandola. The Classics, not *vs* 'the moderns' . . . but their place in a plenum containing XIXth century Europe, the Orient, prehistoric art, Africa, etc. In short, a *full* culture, with cinema and modern mechanics. Not merely overawed by high sounding reputations nor squashed by disbelief in the past.[2]

Between America where "they" know nothing and continually discover the moon, and Germany "where they know everything and make no distinction between anything and anything else" one might aspire toward

Guide to Kulchur summarizes Pound's approaches to a "reading program." Yet, rather than presenting the cultural plenum (which would, after all, mean a hermetic and completely full space in which all culture is complete and further movement and change are impossible, the very opposite of his ideal of metamorphic culture), it catalogues his various provisional methods for "making it new" and opening the cultural heritage to future transformations. From one perspective, it is a book at odds with itself; it is at least subversive of its own stated aim simply to guide "men who have not been able to afford a university education or . . . young men, whether or not threatened with universities, who want to know more at the age of fifty than I know today" (GK, 6). It clearly refuses to yield simple facts. Instead, Pound, who admits that he wants "to COMMIT" himself "on as many points as possible" (GK, 7), here questions the use and organization of "cultural facts" from several perspectives. He does not offer a single method or program for reading and writing poetry but instead advances a procedure which, like that of all the other reading texts, uses creative works and random impressions as reading strategies against traditional aesthetic programs.

Distinguishing this book from *ABC of Reading* and "How to Read," which allegedly establish a "series or set of measures, standards, voltmeters," Pound says that in *Guide to Kulchur* he is "dealing with a heteroclite set of impressions, I trust human, without their being too bleatingly human" (GK, 208). This is in keeping with the autobiographical character of his "data," especially his references back to the "London Vortex of 1914." From among these impressions he selects—or, what is more interesting, asks the reader to select—interpretive "tools," from sources as ancient as Confucian wisdom or as radically avant-garde as Dadaist art, and the latest anthropological methods for uncovering the new in the antiquarian: "Despite appearances I am not trying to condense the encyclopedia into 200 pages. I am at best trying to provide the average reader with a few tools for dealing with the heteroclite mass of undigested information hurled at him daily" (GK, 23).

Pound's apparent indecision about whether his judgments are "totalitarian" or merely encyclopedic, "definitive" or radically opposed to logical classification and argument, is strategic. Aware that he could not fully master all the suggestions his elliptical works compress, he claims that some day the whole corpus of his prose will be published and thus fill in the lacunae of such guidebooks which, in the meantime, can serve for the more comprehensive and methodical works he holds in reserve, "the half million words written in my attempt to arrive at my conclusions."[4] Yet, with the same gesture that promises system, he opens his text to future generations of editors—but not epigones:

> No living man knows enough to write:
> Part I.   Method
> Part II. Philosophy, the history of thought.
> Part III. History, that is of action.
> Part IV. The arts and civilization.

> Even though what I am about to say might be sorted out under such headings. (GK, 23)

Pound there parodies the conventions of guides and treatises. His book has six parts, not four; these are subdivided into thirteen sections and fifty-eight chapters, which seem to move arbitrarily and asymmetrically from topic to topic. Furthermore, Pound does not follow his four-part scheme—at least not in any literal or linear fashion, though the method and the questioning of method remain on the horizon.

Indeed, while *Guide to Kulchur* does not conclude his series of books on reading and culture, it is a book of conclusions, a résumé of his interests as they evolved from the earliest days of his poetic career to his job as Fascist propagandist. Yet his most successful argument, one which subverts his own political and pedagogic totalitarianism, is not a thesis which extends examples toward a conclusion, but a bundle of fragments, each of which stands as the final judgment in deliberations and confrontations which took place before the writing of the book. Thus, spilling over into the unpublished, uncollected, even unwritten prose, his text assaults "monolinear syllogistic arrangement" (J/M, 28), which is the main target of the later, rhetorical prose.[5] Nevertheless, Pound claims for his words the authority and coherence of legal and grammatical *sentences:* "The following summary must be understood in relation to the uncollected half million words of exposition, just as the sentence in a law case must be understood more or less in relation to the court proceedings which have previously taken place."[6] He ironically grounds the logic or truth value of his text on a synecdoche; this figure, on which all examples and conclusions depend, can hardly be stabilized in his "kinematic" text. Despite his ex cathedra pronouncements, then, he does not judge modernism as a representative artist; rather, he interrogates the categories by which modern art was being incorporated into the traditional curriculum by Eliot and others.

The status of his text and the authority of his opinions become part of an ongoing polemic, when, perhaps with an eye to Eliot's roles as publisher and university lecturer, Pound vows to escape the "new academicism," to avoid founding a system along the old (pedagogic and/or genealogical) lines:

> Were I to make a small anthology of the principles exposed in my program I shd be disquietingly aware of a danger, the danger of creating a new academicism. If that program was taken as the program of the old curriculum has been taken for so long this danger wd. be very real. It is not expedient that the student shd accept my opinion about any given passage.[7]

Notwithstanding his obtrusive—and deliberately(?) ironic—self-effacements, he no doubt intends students to take his opinions seriously. His seriousness is manifest as much in the rhetorical tropes—that verbal play anathema to sober argument—which he discovers and reinscribes in philosophy as it is in the angry *rhetoric* (not only tropes but also polemics) he directs at usurers and the other misusers of culture. Therefore, when

we turn to his unconventional readings of Western metaphysics and Confucianism, his attacks upon a Western tradition said to begin in Plato's
Academy, we find that they involve him in German anthropology even
as they repeat Pound's earlier interest in the interpretative nature of the
Chinese (written and moral) character. His own borrowings from other
systems create an inflationary spiral of interpretation, allusion, and abusive citation. If he does not misuse, he at least overuses his various cultural sources.

By continually shifting perspectives and terminology, Pound interrogates distinctions which had seemed dead letters, dead issues, and dead
metaphors. For instance, by tracing the definitions of "culture"—which
various scholars have translated, though without agreeing on a singular
definition, as "learning," "civilization," "*paideia*" and "Kultur"—back
to the pre-Socratics and from America to Africa, he unsettles the fixed
idea of Western classical culture. And by attending to those actions and
inventions which overlap art and philosophy, science and politics, he
frustrates not only his own methodological ambitions but also the claim
of any one method to master the play of discourses and languages he
finally calls Kulchur. At one point, for instance, he claims that real method
underlies this intertextual madness, though even here he assigns it an
untranslatable name, an old coinage, indeed a catachresis:

> I mean to say the purpose of the writing is to *reveal the subject*. The *ideo-
> grammic method* consists of *presenting one facet and then another* until at
> some point one gets off the dead and desensitized surface of the reader's
> mind, onto a part that will register. *The "new" angle being new to the
> reader who cannot always be the same reader*. (GK, 51; emphasis added)

He looks for an ideogram which will also perform the chemical or surgical process of cutting through old associations. "Ideogrammic" is asked
to stand for more than one of his old enthusiasms and dicta. In such a
"system," it is no surprise that the "subject"—whether in the sense of
the grammatical subject who wrote the *Guide* or the topic (subject matter)
of his writing—cannot always be the same. This shaky framework characterizes his effort at once to inscribe a pattern on his own varied cultural
experiences and to gather together key texts in flexible new configurations. Clearly, the reader of Pound's text must also adopt a strategy for
patterning its open-ended interpretive possibilities. It becomes necessary
to examine his notion of Kulchur and its varying contexts in the evolving
discourse of contemporaneous German cultural history. Even this limited
task involves questioning Pound's reinscription of Kulchur into the philosophical tradition which had banished its anarchical force.[8] Therefore,
Plato and Aristotle, whom Pound treats with uncharacteristic deference,

require at least passing comment. By way of following Pound's own in-direction, if along but one of his "new angles," we will come round to *The Pisan Cantos* which repeats several of the topics and strategies of his Kulchur.

II

With the ultimate goal of setting ideas into action, and texts into play, *Guide to Kulchur* steers a course between "German" abstraction and "Chinese" or ideogrammic detail. In the terms of this chapter's third epi-graph, Pound attempts to infuse American culture with a "living philos-ophy"—a phrase he does not consider oxymoronic. In other words, he does not want to fill America's ideological vacuity with empty abstrac-tions, or to write a book of American aesthetic thought, but to advance ideas which will motivate action or generate new wisdom out of ancient beauty. He proposes, then, to read the oldest philosophers in a new light: "The student knows, or can ferret out the evidence, that Zeno, Epicurus, Pythagoras did teach a modus vivendi, did advocate modes of life, and did not merely argue about certain abstractions" (GK, 25). In this regard, the Confucian text, grounded in experience and issuing in proper conduct, is exemplary: "He [Confucius] demanded or commended *a type of per-ception,* a kind *of transmission of knowledge* obtainable only from such concrete manifestation" (GK, 28; emphasis added). Even here Pound stresses the movement and mediation, the reception and communication, not the finality of ideas and facts. Just as he discovered Social Credit economics in the *Divine Comedy,* he finds economic, even ecological data in the work of Leo Frobenius, the German whose "prospective ar-cheology" becomes, like Mussolini's draining of the swamps, an act of will and a value judgment against "mere philosophy."[9] Thus, by a fa-vorite trope of substituting prehistoric art for modern science, Pound in-sists that the most ancient texts and esoteric readings store a "double charge" of new (or renewable) ideas like those artifacts and artistic de-signs in which Frobenius uncovered a functionalism that not only served an ancient culture but offered a method of discovery which opened onto the future, making itself new: "'Where we found these rock drawings, there was always water within six feet of the surface.' That kind of re-search goes not only into past and forgotten life, but points to tomorrow's water supply. This is not *mere* utilitarianism, it is a double charge, a sense of two sets of values and their relation" (GK, 57).

Whether in the German or Chinese (both macaronic at best), Pound addresses commonplace actions and opinions as well as the reified art and ideas of the great tradition. Still, his translations and transpositions of heterogeneous quotations do not comprise an ethical or aesthetic system. Instead, they agitate, recirculate, and otherwise "activate" out-worn words and ideas, but ideas which, he claims, retain the potential for action:

> We must make a clear cut between two kinds of "ideas." Ideas which exist and/or are discussed in a species of vacuum, which are as it were toys of the intellect, and ideas which are intended to "go into action," or to guide action and serve us as rules (and/or measures of) conduct. (GK, 34)

Pound's emphasis on conduct, his preference for the active, as against the contemplative or aesthetic life, perhaps informs his choice of texts for analysis. For example, he considers *The Nicomachean Ethics,* which occupies the last chapters of his *Guide,* not a philosophical treatise but a practical handbook for the education of a young citizen. It was, in fact, addressed to Aristotle's natural son, Nicomachus; similarly, *The Confucian Analects,* with which Pound opens his *Guide,* prescribes the steps to good character, addressing topics as diverse as music and archery.[10] Nevertheless, Pound fails to guide the reader through these ethical treatises or to extract from them nostrums or truisms to live by. He leaves conduct to individual judgment, which is to say, to the reader's engagement of the text—to the act of reading. Recalling his title, he jokingly explains his reluctance to offer final interpretations or univocal advice: "Contract calls for a guide TO not THROUGH human culture. Everyman must get to the insides or the inside of it for himself" (GK, 343). This focus on prepositions (or pre-positions) seems altogether appropriate, since Pound is more interested in disclosing the errors of philosophical *positions,* or the fundamental ideas that comprise Western culture's metaphysical *predisposition,* than in building a new ethical or aesthetic system on the old assumptions.[11]

Thus, Pound's topics in *Guide to Kulchur,* as in his other readers, include "active ideas," "moving images," "kinematic" sculpture, music, poetry, and "Paideuma," which encompasses all these, but not action itself. Pound implies that, unlike the "thing in itself" (*das Ding-an-sich*), action, because it occurs in time, does not admit of philosophical analyses or the translation into system. He turns this apparent inability of metaphysics to deal with history into an argument for the superiority of an art closely affiliated with the "verbal" thought of the East and of primitive tribes and hence with the economy of living in those cultures. Typically, however, he does not completely divorce philosophy or abstract thinking from "poetic thought"; instead, by playing Confucius off against Frobenius, he revises and supplements the philosophical tradition in such a way as to show that, from its beginnings, philosophy has always been troubled by—even inhabited by—poetry and poetic language, and that philosophy had weakened itself by abandoning or repressing the vitality of tropic thought for the rigor (and rigor mortis) of abstraction.

To this end, Pound introduces the Confucian "New Learning" and Frobenius's anthropology near the beginning of his own text, in order to draw upon them as measuring rods or relatively fixed interpretations of the whole of Western metaphysics from Heraclitus to, say, T. S. Eliot. Not only does Pound fail—"refuse" is more appropriate for his deliberate "nonmastery"—to resolve his own ambivalence with regard to the relative value(s) of generalization and the accumulation of detail, theory and praxis, "poetry" and "science" or "philosophy," he discovers that these same conflicting allegiances recur throughout the text of philosophy. For instance, he takes a certain pleasure in disclosing that "Arry Stotl," another of his irreverent (mis-)spellings, had difficulty in choosing between two ethical systems, one based on the enforcement of moral categories (*The Nicomachean Ethics*) and the other based on right living and thus the search for happiness and beauty (*The Eudaemonic Ethics*):

> The weakness is dyed in the very contrast between two juxtaposed words, the old thaumaturgic EUDAIMONIA and the particularly urban THEORIA. Arry is hung up after one age and before the swings and round-abouts of the nine heavens came into pan-European favour. He wd. have had such a good time 12 or 15 centuries later. (GK, 340)[12]

Indeed these two Greek words, *Eudaimonia* and *Theoria,* which might be thought to indicate Aristotle's crucial intervention in an already old argument, are at the base of Pound's own quandary about whether to prescribe the right interpretations or merely to present such translative moments in the philosophical and poetic tradition(s). Moreover, as we shall see, in thus com-*pli*-cating his definitions and examples of Kulchur, Pound privileges those scientific and philosophical methods originating in poetry, myth, and even rhetoric—as in the passing allusion to Gorgias's conquest of Platonism and his recommendation of the earliest Confucian text, *The Odes,* a philosophical poem marginal to the ethical works.

The "New Learning" of the *first* beginning of *Guide to Kulchur* (I say "first beginning" because, like *The Cantos,* this text has at least two beginnings, launched, as it were, on two radically different traditions, one representing ancient facts, the other a pattern of transformations) promises to correct philosophical abstractions by accumulating details which the subsequent text will presumably organize. But how? Relying on his familiar pictographic hunches he merely presents the Chinese "sign for learning"; as though his chosen nonacademic audience can read it:

The dominant element in *the sign for learning* in the love of learning chapter *is a mortar*. That is, the *knowledge must be ground into fine powder*

(GK, 21; emphasis added)

This imperative, which presumably supports the claim that "all knowledge is a rain of factual atoms" (GK, 98), becomes programmatic, as Pound heaps up apparently random details from newspaper clippings, concert program notes, and fragments drawn Dada-like from texts as various as the pre-Socratic adagia and modernist experiments, especially his own earlier speculations which advance philosophemes poetically, or offer poetry as a "rag-bag"—if not Malatesta's "postal bag" of ideas. For example, he cross references his enthusiasm for *The Analects* with his interest in Fenollosa's readings of Western grammar and logic, allowing his rather idiosyncratic Confucius to intervene in the modern sciences:

> Ernest Fenollosa attacked, quite rightly, a great weakness in western ratiocination. He pointed out that the material sciences, biology, chemistry, examined collections of fact, phenomena, specimens, and *gathered general equations of real knowledge* from them, even *though the observed data had no syllogistic connection.* (GK, 27–28; emphasis added)

The opposition of "real knowledge" to "syllogistic connection," a familiar Poundian juxtaposition as well as a stylistic principle, was already there in Fenollosa's notion of the verbal, metaphoric, and metamorphic force of Chinese writing, set against Western logic—especially the Christian *Logos*. But rather than pursue this line of inquiry in Fenollosa's terms or with his data, Pound borrows Frobenius's analyses of the even more ancient African languages and culture, and the German's ready-made "kinematic"—antisyllogistic—connections between linguistic and social phenomena. It mattered little to Pound that Frobenius's work departed radically from the facticity Pound had attributed to *The Analects;* in fact, he used this contradiction to affirm both sides of the oldest metaphysical schism, the differences between Heraclitus and Democritus.

Frobenius's anthropology was compatible with Pound's encyclopedic as well as his theoretical tendencies. Furthermore, because it addressed the verbal—or active—and concrete aspects of primitive languages, *Kulturmorphologie* provided a model, even a style, for mediating between scientific and poetic writing. In this vein, Pound discovers the ideogrammic method already at play in the works of a French ethnologist working, however differently from Frobenius, on African language:

> Levy-Bruhl points out the savage's lack of power to generalize. He has forty verbs where we have two or three verbs and some adverbs. What Levy-Bruhl says about the verbs of savages, what Fenollosa says about verbs in Chinese, what I had written about Dante's verbs before I had Fenollosa all joins up. (GK, 107)[13]

Such fragmentary notes or palimpsests of self-reflection do not, however, delineate a tradition or a method. Pound is more than content to leave them as catalogues of his favorite metaphors (of metaphor). Yet Frobenius offers a sort of "master metaphor"—or at least a key figure for reading several modern artistic and cultural movements back into the tradition.

"Paideuma" was Frobenius's coinage for the outworn notion of *culture*—or *Kultur*, the German word, which had yet to undergo its full expansion (or better, rigidification) into the National Socialist notion of (*volkshaft*) *Kultur*.[14] When Pound borrowed this already problematic notion, he modified it to fit his other favorite schemes. Therefore, we might look at his reading of (his expectations for) paideuma before tracing its workings back to the system—really Spengler's as much as Frobenius's—from which it came. Pound's clearest definition of paideuma reinscribes his own notion of an inherently organic yet critical poetics into his concern for clear and distinct ideas or, in other words, reintroduces the notion of poetry as a kind of critical (active) thought necessary to all Idea and philosophical formulation. Here he fixes on a particularly difficult self-reflexive moment in the Frobenian text which he nevertheless cites as proof of its rigor:

> To escape a word or set of words loaded up with dead association [culture and civilization] Frobenius uses the term Paideuma for the tangle or complex of the inrooted ideas of any period. (GK, 57)

Following this passage, and as though prompted by Frobenius's lexical quandaries, Pound commences *Guide to Kulchur* all over again, beginning with its title. Neither "The New Learning" nor "Paideuma" would have been well received by his readers, but he claims to have intended both when he settled for "Kulchur": "Even were I to call this book the New Learning I shd. at least make a bow to Frobenius. I have eschewed his term almost for the sole reason that the normal anglo-saxon loathes a high sounding word, especially a greek word unfamiliar" (GK, 57–58). Thus, to avoid neologism and/or translation, Pound ironically chose

"Kulchur," an *Amurikan* word which falls between German *Kultur* and English culture. Perhaps his coinage marks the phonetic spelling for the sloppy vowels and guttural consonants of certain American dialects and thus represents a gesture at the "real speech" of the people. In any case, more than once, he raises questions about his key term only to leave the answers open, like all his readings, to further interpretation: "GUIDE TO KULCHUR. Ridiculous title, stunt piece. Challenge? Guide, ought to mean help other fellow to get there. Ought one turn up one's nose? Trial shots" (GK, 183).[15] Such instability—or irrepressible translatability—of his terminology is in keeping with the movements, the turning, of paideuma.

Continuing self-consciously to reflect on his appropriation of Frobenius's notion, Pound calls attention to its flexibility. In fact, paideuma was already there in his own early attempts to save Western civilization; just as it can be traced back to the origins of "natural"—also "cultured"— man: "When I said I wanted a new civilization, I think I cd have used Frobenius' term. At any rate, for my own use and for the duration of this treatise I shall use Paideuma for *the gristly roots of ideas that are in action*" (GK, 58; emphasis added). As a name for the various strata of civilizations and styles which had inhabited the same locus over the course of millennia only to become an excavation site for archeologists who might nevertheless find "living artifacts," paideuma is also a figure for Pound's reading procedure—and for *The Cantos*. It is almost another word for trope. Perhaps not accidentally, then, his reading—hardly a literal translation or interpretation—of Frobenius recalls *The Chinese Written Character*'s description of Chinese characters as "valid scientific thought," which "entails following as closely as may be to the entangled lines of force that pulse through things" (CWC, 12). Pound is, after all, establishing the scientific—especially in the sense of "experimental"—status of his criticism. Such references back to earlier notions and forward to a comprehensive theory or poem yet to come enable Pound to order his own experiences within the energetic vortices—not only the London Vortex—of modern culture. Thus, he simply reads (or reinscribes) Frobenius into his own archeological speculations, which had covered some of the same ground—at least touching on the topics of cave painting and the educational value of knowing the capitals and museums of Europe. This double revision, in which Pound's "autobiography" undergoes changes as stunning as those to which he subjects the words of the German anthropologist, puts ideas into action retrospectively and thereby disrupts the established order of literary history and influence.

He insists, both indirectly and explicitly, that Frobenius's work is "prospective," "not retro-spective" (GK, 57), recalling Emerson's condemnation of American thought as retrospective in the opening sentence of *Nature:* "Our age is retrospective. It builds on the sepulchres of the fathers."[16] And, risking the dread Nietzschean disease of "perspectivism," he applies this principle to a speculative reading economy: "The 'New Learning' under the ideogram of the mortar can imply whatever men of my generation can offer our successors as means to the new comprehension" (GK, 58). Paideuma and the New Learning signify a certain randomness which parodies tradition, and advocates a new notion of language (as play) that tropes or transforms received ideas into instruments of critical thought. In this regard, Pound works within, even as he rubs against, an American grain set forth in Emersonian rhetoric. Like the first "American Scholar," Pound insists on the new and the revolutionary, on "making it new" by going behind and before culture to a nature that is somehow futuristic. Frobenius, an anthropologist and Africanist fascinated by the kinematic and the cinematic, was an irresistible choice to turn to for ways of indexing the primordial and the avant-garde.

Since Frobenius is hardly the influence Pound anticipated, a brief account of his work is in order here, especially since he was another Poundian culture hero and not merely an archivist, as Eliot claimed. "The value of Leo Frobenius to civilization," Pound insists, "is not for the rightness or wrongness of this opinion or that opinion but for the kind of thinking he does" (GK, 57). Like Gourmont, he represents a manner of discovering within the data he analyzes a truth, in the form of patterned energies, of Pound's own thought and the rightness, or direction, of his cultural project: "He has in especial seen and marked out a kind of knowing, the difference between knowledge that has to be acquired by particular effort and knowing that is in people, 'in the air'" (GK, 57). Thus, the African folktales and cave drawings as well as Frobenius's insights about these are part of the "cultural instinct," an oxymoron Pound learned from Gourmont. Biographical accidents might also have made the German anthropologist seem to reflect Pound's own disaffection with traditional discourses. As Pound had abandoned his doctoral studies for more direct research into his poetic origins, Frobenius, inspired by the discovery of the Paleolithic cave paintings at Lascaux, left his degree unfinished in order to trace world culture back to its origin—or ur-forms—in Africa. Based on Spengler's new discipline of *Kulturmorphologie,* Frobenius's research was a critique of modern science and aesthetics, which he saw as abstract, mechanistic, and ethnocentric. Despite the fact that Pound made him into a "scientist," or at least situated him in a (revolutionary) patheon of those poet-scientist-inventors who founded systems, Frobenius's goal was to criticize scientism—not to complete the traditional or to found a new science. His privileged form, the oral folktale, nevertheless suggests an onto-theo-logical nostalgia for a speech prior to writing, technology, and science, and hence a language that could give access to primordial and original thought. Thus, he would simply repeat the metaphysical assumptions of the culture—and the methodological procedures of the history—he hoped to go beyond, driving philosophical speculation further eastward and back in time in his search for an originary poetry free of philosophy, a direction that Pound's thinking also at times seems to take. But this is not the whole story, and not the element Pound emphasized in his selections.

Frobenius's most popular work, the only translated part of his massive notes and archives (some of which were assembled into books published by his institute, *Das Institut fur Kulturmorphologie* in Frankfurt), was the collection of African tales and creation myths. An anthology of these, *African Genesis,* was published by Faber & Faber in 1938, the same year as *Guide to Kulchur.* A few years later, when he was imprisoned at Pisa and forced to compose his poem out of the remembered fragments of his personal cultural residuum, Pound transcribed parts of two of these stories into *The Cantos.* Yet Pound was more engaged by Frobenius's research on African cave paintings which, like Fenollosa's Chinese Written Character and Gaudier-Brzeska's "Paleolithic sculptural vortex," suggests the graphic and figurative (in both senses, tropic and morphic) origins of language. This line of Frobenius's research culminated in the first American exhibit of African art—and not without Pound's intercession.[17] Interestingly, it was Iris Barry, one of the first "*les jeunes,*" who organized the exhibit at the recently founded New York Museum of Modern Art. And the exhibit catalogue, written by Frobenius and Douglas C. Fox and published by the museum in 1935, insists on the parallels between "primitive forms" and avant-gardism which Pound had noted as early as 1914, in his Vorticist Manifesto. Frobenius thereby provided data and, if not a theory, an interpretive model for avant-gardists who wished at once to complete the aesthetic tradition and to reverse its "priorities"; that is, to revise both the direction of its progress and the privilege afforded to that art which originated in Hellenic Greece. If Frobenius's "paideuma" had not existed, Pound and his cohorts would have had to create it—and in a sense they did.

Just so, Pound seems to collapse the general procedures of *Kulturmorphologie,* which analyzes the "life-cycles" of biological as well as linguistic forms (*morphē*), into the notion of "paideuma," which thus covers names of the movements within and between recurrent cultural forms as well as the forms themselves. Frobenius defined the term simply as "the spiritual essence (*Geist*) of culture in general."[18] Pound would not of course settle for anything as abstract and German (even Hegelian) as "essence" or "spirit." And, as it happens, his more metaphoric or even metamorphic definitions seem appropriate. Even if Frobenius would not have approved all Pound's revisions, we may still conclude that "inrooted ideas of any period" and "gristly roots of ideas in action," phrases cited above, capture the interpretive force and organicism of his model.[19] Therefore, despite his free translation of its key terms, Pound manages to stay quite close to Frobenius's text. For example, while he seems to have mistaken them for "scientific method" and "proper naming," ostensibly his chief criteria for pedagogic and theoretical acceptability, he nearly repeats the terms of this programmatic statement:

> *What I have to say does not answer the question "This is how things are,"* but rather *"This is how they are to be understood."* My investigation is not concerned with modern psychology or physiology: it is obliged to pursue its own way. . . . At the same time, *it is a modest enquiry,* and in a sense an *unscientific one. This may be perceived in the use of special terminology* and forms of language which are not readily intelligible, because their subject matter is not. In particular, I have found it necessary for certain purposes to *replace the word "culture" by the special term "Paideuma."* (Frob, 21; emphasis added)

We have already noted Pound's similar self-consciousness about "Kulchur." And, while Frobenius does not explicate the Greek etymology of "Paideuma," Pound at least alludes to it: "a greek word unfamiliar. . . . As I understand it, Frobenius has seized a word not current for the express purpose of scraping off the barnacles and 'atmosphere' of a long-used term" (GK, 58). The long used term is "culture," and, if he does not mark the steps of his argument (really his Gourmontian dissociation), Pound traces it back to its Greek roots in *paideia,* or culture. In this vein, with Frobenius but one of his precursors, Pound delineates a sort of countertradition which abolishes the constraints Plato had enforced on *paideia,* when, in the effort to banish poetry (and Rhetoric) from philosophy, he cautioned against *paidia,* the Greek word for the "play" associated with sophistry, nonserious or figurative language, and dramatic poetry. Plato's argument can be found in the *Gorgias* as well as *The Republic.*[20] Although Pound cites neither Platonic text, his revisionary history of philosophy focuses on such things as Aristotle's ambiguous ethical terminology as key points—inventions (rhetorical *inventio*) and translations— in his "philosophical paideuma." This reading of differential terms can be related to Frobenius's "kinematic maps" of changes in the Paideuma—

which, somewhat in contrast to Pound's usage, are translatable and continuous. Yet Frobenius's analyses of African folktales also emphasize the tropes and other variations which depart from the recurrent themes and "oral formulaic" markers of the tales. In this way, a paideuma (or manifestation of culture) is characterized more by its differences from than its repetitions of an underlying *Paideuma* (Culture). For Pound, paideuma is plural and transformative—or figurative. It is a trope of trope closely allied to his notion of poetic language as a critical and metamorphic force playing through traditional cultural and aesthetic forms, opening them to new arrangements.

Finally, to return to Frobenius, *Kulturmorphologie* does not address "things"—or the text—directly ("how things are"); instead, it ex-*pli*-cates the various layers of ideological and linguistic mediation that comprise modern culture and its discourses.[21] While it never approached Nietzsche's rigorous deconstruction of the philosophical *Darstellung,* or Foucault's critique of the *episteme,* in a work suggestively titled *The Archaeology of Knowledge,* Frobenius's anthropology, like Gourmont's physiology, can be seen as a sustained polemic against the rationalism, moralism, and ethnocentrism inherited from a narrow or philosophical version of classicism.[22] For instance, Frobenius opposes the "intuitive principle" (which he saw as fundamental to both primitive thought and morphology) to the mechanistic procedures of nineteenth-century science and philosophy. Adumbrating more than one Poundian theme, he suggests that anthropological research demands involvement, action, and the ability to read symptoms and forms in the chaos of modern reality without turning them into sterile taxonomies:

> *The mechanistic principle is, like a railroad track, the shortest means of reaching a given end,* but it prevents us from taking a broad survey of the country as we pass through it. . . . It is a type of biological or psychological approach based on albumen tests, laws of association, motives and impulses, ganglion cells and nerve tissues, all duly classified and reduced to dry formulae. *The intuitive observer by contrast, seeks to enter with his whole being into the lawless profusion of spiritual activity,* at the same time distinguishing the significant from the trivial, the expressive from the merely accidental. . . . *Instead of petrified laws and formulae, he discovers symbolic events and types of living, breathing reality.* (Frob, 19; emphasis added)

This passage condenses several nineteenth-century methodological issues that had earlier made their way into Pound's text: the value of unsystematic observation and chance discovery, the inadequacy of rigid classifications and formulae to biological and historical (or "living") data, and the hope of replacing endless detail with "symbolic events." Pound names these symbolic events "luminous details" and "ideograms," using the latter, in a sense that goes beyond Frobenius to encompass his earlier definition of the Image as a "complex" of emotion and ideas in an instant of time. We can trace Pound's interests through the texture of contemporaneous German historiography—which was involved in the new(ly named) discourses of anthropology, archeology, and philology. This new science, which almost no one followed to the letter (and certainly not someone with Pound's slight grasp of the German language and tradition), had a profound influence on modern art and aesthetics.

III

Pound's attempt to redefine "culture," to gather its "luminous details" into a literary pedagogy and ultimately into a new political order, brought him, by way of Frobenius and Burckhardt, into conjunction with modern German (if not neo-Nietzschean) redefinitions of *Kultur*. Thus, his "Kulchur" differs radically from the Arnoldian—which is to say "English" but also "Eliotic"—notion of culture as a stay against the anarchy of modern urbanization and the attendant perils of foreign influence and the "philistinism" of the new bourgeoise. The English tradition—the tradition which narrowed the category of culture—tended to conserve the values and forms of British poetry and religion against primitivism on the one hand, and Futurism or more generalized avant-gardism on the other hand. We have already considered Pound's journalistic attacks on this "Anglican," "Classicist," "Royalist" position represented by Eliot. Pound's counterposition, it is obvious, did not come from the air, especially not the English air. Despite his frequent slams at German thought and any sort of philosophy, he felt obliged to treat both in some detail in *Guide to Kulchur*. From his recollections and his *newer* methods, it is clear that, in addition to the "Nietzscheanism" which he found translated through Gourmont, as early as 1911, in his "New Method of Scholarship," he had relied on the more purely German perspectives of Jacob Burckhardt, a Nietzschean precursor if not a Nietzschean. Even in the earlier reading texts, which seem directly opposed to Germanic philology and the German university system, Pound draws from Burckhardt and perhaps from Spengler, both of whom address the "force" of ideas which at once constitutes and disturbs *Kultur*.[23]

Burckhardt's name is now synonymous with an old-fashioned notion of the Renaissance, but his interpretation of thirteenth- to fifteenth-century Italian culture was part of a nineteenth-century discursive revolution which, in light of Hegel's world-historical Mind, linked traditional history (that is, biographical and martial "facts") with aesthetics, philology, epistemology, politics, and economics. *The Civilization of the Renaissance in Italy* (1878), his major contribution, was ostensibly one of Pound's sourcebooks for readings of the Late Medieval papacy, the *condottieri,* various artist-scholars, and Dante (as political poet). While he did not second Burckhardt's condemnation of Malatesta's bloodthirstiness, Pound apparently borrowed details about *The Tempio,* including the burial of its artists and architects inside their creation and its pagan rededication to Isotta—though, as usual, without acknowledging his source. As already noted, he drew "luminous details" from Burckhardt's analyses of Venetian commerce: "When in Burckhardt we came upon a passage: 'In this year the Venetians refused to make war upon the Milanese because they held that any war between buyer and seller must prove profitable to neither,' we come upon a portent, the old order changes, one conception of war and of the State begins to decline. The Middle Ages imperceptibly give ground to the Renaissance" (SP, 22). One could no doubt trace other Poundian details back to Burckhardt. What is more important, the German historian, however unaccountably, contributed to Pound's notion of culture as an inherently anarchical force which could—paradoxically—inscribe the "right hierarchy."

Burckhardt articulated his cultural theories in lectures delivered at the rather isolated University of Basel in the 1870s. If Nietzsche heard his words, Burckhardt's notes were certainly unavailable to Pound, since they remained unpublished until 1940 and untranslated until 1943. Nevertheless, they outline a "program" compatible with *Guide to Kulchur* and *Will to Power*—that is, they are a compendium of topics and details which cannot be reduced to a narrative, or to a total history, but function as *topoi* or even themes which suggest a cultural pattern. Pound's discovery (as early as 1911) in Burckhardt's *The Civilization of the Renaissance in Italy* of types and tropes which function as "luminous details" suggests a method that we can now see was behind the German's corpus all along. For example, here is Burckhardt on the notion that the "facts" of culture are tropes or mental events signifying transformations: "Culture may be defined as the sum total of those mental developments which take place simultaneously and lay no claim to compulsive authority."[24] He explains that culture modifies the two historical constants, the state and religion, and thereby affects the organic process of civilization and its decline:

> Culture is . . . *that million-fold process by which the spontaneous, unthinking activity of the race is transformed into considered action,* or in-

deed, at its last and final stage, in science and especially philosophy, into pure thought.

Its total external form, however, as distinguished from the state and religion, is society in its broadest sense.

Each one of its elements has, like the state and religion, its *growth, bloom,* and *decay,* and its perpetuation in a general tradition (in so far as it is capable and worthy of it). Countless elements also subsist in the unconscious as *an acquisition bequethed to mankind perhaps by some forgotten people.*[25] (emphasis added)

Pound does not fully endorse Burckhardt's teleology and the privilege afforded "pure thought," but he does imply the notions of Kulchur as "process" and as an "unconscious inheritance" or residuum, as well as (Frobenius's) research into the forgotten origins of civilization, in his compendium of cultural themes, which begins "at the end of philosophy," as it were, "when one HAS 'forgotten-what-book'" (GK, 134). Furthermore, as though reading Burckhardt against himself, Pound employs the phrase "transformation of ideas into action" for his critique of abstract philosophy which, through an initial error, rejected force, energy, and historicity in favor of reified or purified Idea. Just so, by Pound's account, historiography becomes rigid, even inorganic, when it abandons the "action" of history and its makers for their ideas and ideologies: "Run your eye along the margin of history and you will observe great waves, sweeping movements and triumphs which fall when their ideology petrifies" (GK, 52). He does not hereby simply reject the metaphysical tradition; he still hopes to revive—and revise—it for the common reader who views philosophy as "a highbrow study, something cut off both from life and from wisdom" (GK, 24). For instance, by way of a different strain of Platonism, he retranslates the traditional Platonic—also Hegelian and Burckhardtian—"Mind into action," and therewith makes Plato an actor, as it were, in a dramatic history of ideas played out upon the stage of general culture:

If Plato's ideas were the paradigms of reality in Plato's personal thought, their transmutation into phenomena takes us into the unknown. What we can assert is that Plato periodically caused enthusiasm among his disciples. . . .

The history of a culture is the history of ideas going into action. Whatever the platonists or other mystics have felt, they have been possessed sporadically and spasmodically of energies measurable in speech and in action, long before modern physicians were measuring the electric waves of the brains of pathological subjects. (GK, 44–45; emphasis added)

There and throughout *Guide to Kulchur,* Pound is more interested in the movement of ideas—even in philosophical and artistic "movements"—

than in the discrete acts of "factive personalities" or seminal thinkers. Thus, instead of merely compiling lists of examples, or even of "highly charged language," in the manner of *ABC of Reading,* he stresses the need for a history and pedagogy adequate to the cultural process:

> You can't in observing the process fix the exact point where the study pertains to philosophy, that is where one is studying thought and where one begins dealing with history (action). (GK, 43)

Although the concept of history as a biological process is never explicitly advanced by Burckhardt in the way it would be systematically formulated by Spengler, the source of Frobenius's anthropology and by indirection of Pound's notion of Kulchur, we can say that Pound virtually takes this germinal thought of cultural historians as a sort of "luminous detail" by which to characterize modern culture. Thus, historiography would not simply record the facts of a culture but would in effect generate them, leading to his assumption that culture is not a thing itself but a transformational force. Spengler, who remains marginal both to history and to philosophy, attempted to chart such transformations of cultural forms through the whole of human civilization. Notwithstanding the fact that Pound depreciated his generalizations in contrast to the details of Frobenius's African fieldwork, Spengler's writings are a lengthy polemic against philosophical abstraction and exclusivity.[26] Indeed, in the Introduction to *The Decline of the West* (*Der Untergang des Abendlandes,* 1918), Spengler goes so far as to characterize *Kulturmorphologie* as a repetition of the "Copernican revolution," as a decentering of culture from Greek art and philosophy, if not the overthrow of metaphysics by poetry:

> [Ptolemaic is] the most appropriate designation for the current Western European scheme of history in which the great cultures are made to follow orbits around us as the presumed center of world happenings in the *Ptolemaic system* of history). The system that is put forward in this work in place of it I regard as the *Copernican discovery* in the historical sphere, in that *it admits no privileged position to the classical or the Western Culture* [emphasis added] as against the cultures of India, Babylon, China, Egypt, the Arabs, Mexico—*separate worlds of dynamic being* [emphasis added] which in point of mass count for just as much in the general picture of history as the Classical, while frequently surpassing it in point of spiritual greatness and soaring power.[27]

In passing, we might note that Spengler likens his antisystem's shifty denial of the existence of a center to the Copernican system, which merely replaced one center, the earth, with another, the sun. Spengler's apparent oversight not only proves the dangers of analogy but also the unremitting (self-deconstructive) rigor of metaphors of metaphor and the ubiquity of the sun as a figure for philosophy. However critical he appears, Spengler does not avoid assigning Germany the decisive role in modern civilization. And, as is well known, his history was easily turned into the centerpiece, if you will, of Nazi arguments about the antiquity and purity of the Aryan race—all the way back to India and the proto-Indo-European language. But, like the *reichsdeutsch* interpretations of Nietzsche, that is part of a different figuration and polemic which Spengler himself criticized in the generally ignored writings after *The Decline*.[28]

More important here is Spengler's decision to use figurative language in a scientific or critical discourse. He questions the traditional position of Greece as origin and acme of culture and replaces that classical and linear genealogy with a series of dynamic centers (or microcosms). As ideas, his "innovations" are virtually commonplace, as old as Heraclitus ("Everything flows" [GK, 31, *et passim*]), about whom Spengler wrote his doctoral thesis. What is new or inventive is his substitution of analogies and other rhetorical figures for clear definitions and logical argument, not to say his oracular, even combative, style which has made him an embarrassment to "objective" scholars—much as Pound's indecorous name-calling and out-of-hand dismissals of the literati alienated him from Eliot's circle.[29] When Spengler defined *Kulturmorphologie* he was fully aware of assaulting the German academic establishment, of opening science and philosophy to the critical and sceptical forces of a certain poetic countertradition. He states his debt simply: "Goethe gave me method, Nietzsche the questioning faculty" (Decline, xiv).

Indeed, Goethe was Spengler's major precursor in his battle against Kantian categories, offering him the strategies and tropes of *Morphologie* which he deployed against "mathematical number" and the "iron laws" of logic. The great German poet-*bricoleur*'s interest in science, symbolized by Faust's overarching curiosity, led him to found morphology, a procedure with applications in both linguistics and philology. Goethe's (re-)coinage as a compound of the Greek oxymoron neatly suggests the double thrust of metaphysics and poetry that inheres in this comparatist method—even the OED credits him with the first usage of the term "morphology" and as well with the somewhat pejorative designation "science." The Greek suffix "*-ologie*" affiliates this method with that objective and syllogistic argumentation which yields truth statements presuming to (scientific) universality and predictability and aspiring ultimately to the metaphysical *Logos*. Pound mentions this suffix in his warning against ever more abstract modern knowledge: "Every word ending in -ology in English implies reading generalities. It implies a shutting off from particulars" (GK, 99). By contrast, *morphē,* the Greek word for "form," retains the sense of phenomenalism and particularity, though its application is so wide that it can pertain to anything with a definable structure—that is, to atoms as well as poetic images, which is another possible translation of *morphē.* Furthermore, as Goethe knew when he chose this word, Morpheus is Ovid's (Latin) name for the god of dreams, that unseen force behind the fantastic transformations within our ordinary dreams and the "deity" presiding over the "meta-morphosis" of the Ovidian masterpiece. Just so, morphology considers the changes in form of biological species and the translation of meaning and linguistic structures across different traditions. And, as we have seen, Golding's *Metamorphoses,* which Pound treated as compendious and irrepressibly (trans-)figurative, in contrast to the more historical epics of Homer, features prominently in *The Cantos* which, by juxtaposing *The Odyssey* and *The Metamorphoses,* interprets or rhymes the forms of two cultures—or, more exactly, at least three: Homer's Greece, Ovid's Rome, Golding's Elizabethan England, and, by extension, Pound's *Amurikan Kulchur.*

However tangled, morphology's two lines of form and logic can also be detected in two of the major discourses dating from the nineteenth century, linguistics and biology—that is, Grimm's philology and Darwin's evolution. These two "sciences" come together in the work of Gourmont and Nietzsche, as well as in that of Spengler and Frobenius. Indeed, nearly all those nineteenth- and early-twentieth-century writers anxious to reverse entropy and the apparent conquest of art by sterile philosophy attempted to bridge the two discourses—whether in a "physiological aesthetics" (Nietzsche and Gourmont) or in a "history of culture's living forms" (Spengler and Frobenius).

Since all these critical thinkers who provided details and background for Pound's notion of Kulchur studied linguistic and biological forms, or used metaphors drawn from biology, we might pause for a moment over the scientific province of morphology. Morphology is that subdepartment of biology which identifies, classifies, compares, and names life forms. In the effort to define characteristic structures, the species and genera of the taxonomies of the animal kingdom, morphologists study the homologies, analogies, metamorphoses, and even anomalies of specimens ranging from cells to groups of animals. Thus, morphology's flexible lexicon can begin to graph evolutionary transformations by giving new names to newly discovered relationships. And, while Darwin's theories are not directly indebted to Goethe, his research parallels this grammar of living forms, which at once culminates and destroys the rigid taxonomic order. In the area of linguistics, where morphology still occupies a niche in structuralism and generative grammar, Goethe is the acknowledged precursor of Jakob Grimm, whose brainchild, "comparative philology," indexed grammatical forms to establish Grimm's Law, which traces the lexical and grammatical variants in what the philologists, true to their ethnocentrism and sexism, called the Languages of Man. And, in anticipation of Frobenius's work on African tales, the Brothers Grimm (Jakob and Wilhelm) identified the common elements in German fairytales. Morphology, therefore, conspires in the metaphysical weave of language and life (culture).

Philology's reliance on the tropological aspect of morphology is perhaps more obvious than the debt of biology, if only because the terminology and data of the former are both comprised of the movements within and between words or *texts* (though Foucault and other recent neo-Nietzschean historians of science would insist that biology is every bit as textual or rhetorical). Thus, it is no surprise that, drawing upon the language of biology and, to some degree, on the yet-to-be-born "science" of sociology, the Grimms employed a whole series of metaphors for the orderly growth of languages out of a (mythical) proto-language: "Indo-European *family* of languages," "*roots* of words," "mother tongue"; "dominant," "strong," "masculine" as opposed to "feminine" verb forms, "conjugation," and the like. In Gourmont's *Natural Philosophy of Love* and Pound's "Postscript," such tropes of trope as "spermatic images" and the "ejaculation of ideas," among others, might be thought to parody such grammatical and genealogical models. Pound's criticism retains a pronounced nostalgia for an originary, poetic language, hence one irreducibly plural and metamorphic. When, following Frobenius's method of *Kulturmorphologie,* he finds Kulchur to be "the inrooted idea of any age," this he immediately translates into the broader context where culture and nature, meaning and image or trope, are inseparable even at the (non-) origin of poetry.

Our brief and entirely too schematic consideration of nineteenth-century morphology is not meant to insinuate, much less detail, a family of sciences or a line of investigation, and certainly not a history of influences. Indeed, it suggests the dispersal of certain metaphors and the pervasiveness of what Spengler called "history on the analogy of living forms"—or *Kulturmorphologie.* To return to *The Decline of the West,* we should note Spengler's insistence that, while cultures are aimless, random, unconnected, and, in fact, uninterpretable from outside, they are *homomorphic*—that is, they correspond in external form but share no underlying structure or common origin. He identifies nine or ten cultures by reading morphological relationships, symptomatic concepts and works, out of the chaos of detail of ordinary life plus whatever historical artifacts are available, without, he hopes, taming that chaos into dull catalogues or overgeneralizations. After all, his goal, not unlike Pound's, is to substitute for German philosophy's "abstract-systematic theories" his own "practical-ethical" interpretations (Decline, 20). He does not claim to have founded a new system or to have given the final word. In this vein, seven years before Spengler finally published his magnum opus, Pound called his own method of luminous detail "not dogma, but a metaphor which I find convenient for expressing certain relations" (SP, 28). Spengler claimed, "what concerns us is not what the historical facts at this or that time *are*, per se, but what they signify by appearing" (Decline, 6). Therefore, for the chronological categories "ancient," "medieval," "modern," he substitutes "symbolic events" and "senses of form" which make a culture into "a coherent conception of life" (Decline, 7); this is not, however, reducible to a "philosophy of life." On this point, Frobenius's related definition of culture is yet clearer: "Culture is a product of the mind working invisibly, and the spiritual is expressed in everyday life more than in any 'intellectual process'" (Frob, 23). Likewise, Spengler reprivileges the three traditionally "high" civilizations or philosophies, though under borrowed, poetic names: "Apollonian" (Classical), "Magian" (Middle Eastern or Hebraic), and "Faustian" (Western Europe). And Frobenius's research broadens the application of *Kulturmorphologie* to African art and thus to forms that more radically question historical methodology and culture traditions. Even in the most primitive drawings Frobenius discovers the same, if purer, sociopolitical and aesthetic—or cultural—forms as those of modern Western Europe.

Thus, *Kulturmorphologie* allows one to make sweeping generalizations by employing discrete sets of detail and, despite any claims to the contrary, to view *Kultur* as a sort of continuum—though this line remains, like Pound's various definitions of Kulchur, merely heuristic and open to interpretation. Spengler's work is riddled with inconsistencies, yet his manner of cross-referencing forms and symbols and analogies, his loose organization of significant (or luminous) details, had an obvious appeal for "archivists" like Frobenius and Pound. In a passage remarkable for cataloging Pound's own favorite details, Spengler traces intersections of the forms of "mathematical sense" and "political sense" across several cultures:

> Between *contrapuntal music and credit economics*, there are deep uni-
> formities. . . . Viewed from this morphological standpoint, even the hum-
> drum facts of politics assume a symbolic, even a metaphysical character,
> and—what has perhaps been impossible hitherto—things such as the Egyp-

tian administrative system, *Classical coinage, analytic geometry,* the cheque, the Suez Canal, *the book printing of the Chinese,* the Prussian Army, and *Roman road engineering* can as symbols be made *uniformly* [emphasis in text] understandable and appreciable. (Decline, 7; emphasis added)

Here, Spengler's topics suggest Pound's claim, recurrent through *Guide to Kulchur* and *The Cantos,* that one can judge the degree—if not the exact percentage—of usury in any given culture by analyzing such symptomatic details (the detail/generalization model or synecdoche is the more accurate figure than symbol for Pound's use of examples) as the clarity of musical notation, the distinctness of line in painting, the number and quality of roads and other public works. Moreover, Pound's examples of what, following Spengler, we might call the proper "economic sense"— or even the "economy of proper form"—include Bach's counterpoint, Roman and Late Medieval coinage, Chinese manuscript production and the modern booktrade, and Mussolini's—if not the "first Caesars's"— road building and swamp draining. Of course, he could have borrowed these details from any number of sources, and he sometimes processes these and other data through other metaphors than those of *Kulturmorphologie.* But even this uncertainty of origin and application tends to confirm Pound's Spenglerian notion of the swift transmission and fragmentation of modern *Kultur,* which becomes accessible to understanding only through the most vivid of rhetorical figures and formulations. One recalls that it was his own version of analytical geometry from which Pound appropriated his metaphor for the poetic image in *Gaudier-Brzeska,* but an analytical geometry most notable for its untranslatability into plane geometry.

To conclude the present examination of Spengler, we might note the methodological claims he made for the overriding analogy, an extended organic metaphor, which enabled his comprehensive and radically ahistorical history. In a series of rhetorical questions, he turns the most commonplace notions of the life-cycle into a rigorous new method, nearly a new philosophy, but one deliberately based on the shaky foundations of figurative language and common sense: "For everything organic the notions of birth, death, youth, age, and lifetime are fundamentals—may not these notions . . . possess a rigorous meaning which no one has yet extracted? In short, is all history founded on biographic [and biological] archetypes?" (Decline, 3). Spengler's favorite motif is the four seasons: within the seasonal changes of a culture, which lend themselves to his neat fold-out charts, he places the acts of great statesmen and the works of representative artists and philosophers. This schematic allows him to violate traditional generic and chronological divisions, to identify certain individuals as "morphological contemporaries," and to make predictions about the destiny of modern culture: Pythagoras and Buddha represent Spring in their respective cultures; Phydias and Bach are Summer; Plato and Kant, Autumn; Caesar and ———? (Spengler's blank), Winter. As though following Pound's recommended procedure of leaving blanks for future readers, Spengler dangerously prefigures the Caesarian leader of Nazi mythology.

But Autumn suggests a more interesting critique; for Spengler, whose feelings about philosophy were ambivalent, says that both the Apollonian and the Faustian cultures experienced their intellectual height in the Autumn. In their works, Aristotle and Plato, as well as Kant and Hegel, arrived on the scene of their respective cultures just after periods of great artistic creativity and immediately before the decline of "culture" into mere "civilization," his word for the dormant phase (Winter) when the "cultural spirit" is awaiting the creation of new forms (Decline, 46). While Spengler did not insist on the analogy between the decadence of Greek and German philosophy, he did adopt Nietzsche's strictures against Socrates (and Platonism) as a vitiation or repression of Dionysian power and creativity. As we have seen, Spengler is quite explicit about his own efforts to inject poetry and rhetoric into philosophy, to re-*verse* as it were, the decline of the West by the infusion of modern art and poetry. They are a "transitional phase" in which, at best, the remnants of a more vigorous past are preserved in the Western megalopolis. This pessimistic vision was advanced at the end of World War I, in the wake of one German historical vision, at about the same time Pound was mourning Gaudier-Brzeska and the 1914 Vortex: "The nineteenth and twentieth centuries, hitherto looked on as the highest point of an ascending straight line of 'world-history,' are in reality a stage of life which may be observed in every Culture that has ripened to its limit. . . . Our time represents a transitional phase" (Decline, 39).[30] In lieu of despair or false hope, Spengler proposes a new, affirmative *skepsis* which, going beyond "Nietzsche's naive relativism," would renounce absolute standpoints and accept the inevitable process of history—the Spring following Winter. His own nostalgia, not to say his onto-theo-logical commitment, prevented him from recognizing this skepticism in, for example, the Futurists and the other avant-gardists whose iconoclasm and fragmentary or mechanical "art" he saw instead as proof of the decline. His judgment of decadent art, in fact, stopped far short of Nietzsche's affirmations—and Pound's.

Pound did not consistently apply the model of cultural life-cycles, though his version of what might be called history's recurrent (and nostalgic) modernisms answers the Spenglerian notion of unilinear decline. For instance, he considers Malatesta a culture hero—if not a (proto-)"pro-spective archeologist"—for employing the best artists in order to save the remnants of both Classical art and his own disappearing culture from usurious collectors as much as from the barbarians. By a rather stunning version of the ahistorical analogy, Pound equates his own efforts on behalf of art with those of Malatesta: "Sigismundo cut his notch. He registered a state of mind, of sensibility, of all-roundness and awareness. He had a little of the best there in Rimini. He had perhaps Zuan Bellini's best bit of painting. He had all he cd. get of Pier della Francesca. *Federigo Urbino was his Amy Lowell*" (GK, 159; emphasis added). If inserting a fragment of the polemics surrounding Imagism into the history of the Papal Wars doesn't strike one as quite absurdist enough, at crucial points in *Guide to Kulchur* Pound uses Dadaism, an avant-garde movement which remains unaccommodated by the academy and unacceptable to Amy Lowell, as a critical tool against both canonical modernism and the philosophical tradition:

> A definite philosophical act or series of acts was performed along in 1916 to '21 by, as I see it, Francis Picabia. If he had any help or stimulus it may have come from Marcel Duchamp. . . . The accepted cliché turned inside out, a, b, c, d; being placed
>
> b, d, c, a,
> c, b, d, a, etc., in each case
>
> expressing as much truth, half truth or quarter, as the original national or political bugwash. (GK, 87–88).

Here Pound seems to ascribe to the Dadaists a key role in grinding knowledge into fragments, if not the "fine powder" represented by (the sign or ideogram of) the mortar. However elliptically, Pound extends the attack on monolinear logic into a political act: associating avant-garde alogicality with his own project of "setting ideas into action," he marries a generally leftist radicalism with Fascism. Thus an antiphilosophical doctrine becomes a philosophical act that overcomes philosophy by draining the Italian swamps and performing other community services.

In a yet more schematic passage, he fractures his own narrative as well as that of the history of philosophy, when he causes Dada—or a certain Dadaist rhetoric—to intervene in Aristotle's sober speculations: "In a sense *the philosophical orbit of the occident* is already defined, European thought was *to continue in a species or cycle of crisis:* grin and bear it; enjoy life; *variants of Gorgias' dadaism*" (GK, 120). Surely such passing remarks, intended more to break through "the desensitized surface of the reader's mind" than to delineate a countertradition, do not comprise anything so grand as *Kulturmorphologie*. Yet Pound nearly repeats Spengler's revolutionary rhetoric, the notions of "cycles" and "crises"—the tropings or revolutions—within cultural history. Inasmuch as these figures can be said to have a source, however, his acknowledged debt is to Frobenius. Before we consider Pound's more sustained attacks upon Plato and Aristotle, in which Frobenius, "the Geheimrat" (a German title meaning "privy counselor" becomes, in Pound's argot, a code pun for master and burrowing animal), appears as a name if not a force, we might pause for a moment over that anthropologist's notion of the intellectual stages of paideuma.

Like Spengler's four seasons, Frobenius's three-part model of the intellectual paideuma, childhood-youth-maturity, applies equally to individuals, discrete national cultures, and the whole of Western civilization. Applied to the other, non-Western and primitive cultures Frobenius investigated, this model perfectly exemplifies the presumed cultural superiority of the West, that is to say, of metaphysical or trinitarian thought. The three stages are not necessarily symmetrical or translatable between cultures or men. Every culture, every thinker, has the potential for undergoing the full course of intellectual development, but some might become arrested at an early stage or skip on to a later one. Frobenius believed, for instance, that African tribesmen possessed a disproportionately large reserve of the moral perfection and heightened imagination he associated with childhood. One the other hand, because of the maturity, or near senility, of the whole Western European "race," a typical modern artist might lack the creative imagination altogether and thus merely reproduce received ideas. These stages do not proceed with linear or dialectical necessity, though Frobenius tends to chart them through a continuous history as a dialectic between "intuition" and "mechanism." They can instead be experienced and measured, like the Medieval "humours," as an excess or deficiency. Nevertheless, there is a qualitative loss of intellectual power and creativity from childhood to maturity, and a parallel decline from the primitive to the modern.[31] The great inventions of modern science come from the sort of spontaneous inspiration that motivates the sudden insights of a child. This, in Frobenian terms, the first stage of the intellectual paideuma, "the daemonic world of childhood," involves that spontaneous "spiritualizing quality" by which children, "savages," and great creative geniuses transform the mundane into an imaginative or totemistic world: "The 'ideal' world of early manhood" produces an "intelligence" characterized by philosophical ideas and individual, as opposed to tribal, art; this is the thought and the period of

Classical art and systematic philosophy. The third stage, called "facts in the world of maturity," involves the accumulation of possessions and the mechanical processing of facts in the practical arts and sciences (Frob, 43–48).

By this account, Western culture has passed through all the stages, having Fallen, as it were, from mythic, intuitive, or Edenic thought through a period of great art and philosophy to the modern emphasis on academic conformism and the circulation of received ideas. Yet this course can be avoided, and the very "decline of the West" averted, by those individuals who, drawing upon the ever-present residuum of earlier and foreign paideumas and/or the resources of their own imaginations, attain an integration of the three types of knowledge. It is possible, Frobenius suggests, that at the exhausted end of Western philosophy a new infusion of (primitive, African, or Eastern) ideas will culminate in a unity of the best of all thought, or the best of all worlds, in "daemonic genius, individual idealism, and intellectual purpose" (Frob, 54). Rather than rejecting or questioning the values of the philosophical tradition, his work, at least if one takes seriously its teleology, completes the system, coming full circle to an original unity. But this is not exactly how Pound read him.

In *Guide to Kulchur,* Frobenius is an innovator who gathers suggestive fragments into new patterns and thus offers Pound details by which he can unsettle such entrenched ideologies or systems as "German thought" and Classical philology. For instance, Pound credits him with a special, if ineffable, "sense" of popular culture, then cross-references this sort of intuition with the recent discoveries of W. H. D. Rouse, the founding editor of the Loeb Classics:

> I reiterate our debt to Frobenius for his sense of the reality in what is held in the general mind. Dr. Rouse found his Aegean sailors still telling yarns from the Odyssey though time had worn out Odysseus' name down through O'ysseus, already latin Ulysses, to current Elias, identified with the prophet. (GK, 79).

Thus Frobenius's research in the primitive or mythic unconscious becomes, for Pound, a way of showing the persistence of the Western textual tradition. And, what is perhaps more disturbing to Frobenius's attempts at broadening the scope of cultural history and anthropology, Pound privileges the dispersal and the descent of poetry from the traditional high point of Greek culture to the gossip of sailors, by assigning it the name Kulchur. Later, he uses the same colloquialized Homer, but Frobenius has been replaced by (an allusion to) Flaubert: "If you have a passion for system, you can try the picture of Moeurs Contemporaines in Homer"

(GK, 260). This collapsing of preclassical and present writing had already been anticipated and even realized in the self-reflexive structure of *The Cantos,* for example, Canto VII, where the narrative "ply over ply," of Flaubert and Henry James, is wrapped upon the earlier metamorphic themes of the ancients, composing Pound's version of *Kulturmorphologie* as a "house of fiction" or, as Pound says, "Rooms, against chronicles." Pound's accounts of the transformations of folk and popular lore into literature and out again tend to be much more disruptive and antisystematic than those of Frobenius, who was always aware of the problems his research posed traditional histories and aesthetics.[32]

For instance, reflecting on his own efforts at translating the wisdom and power of the African tradition into the terms of Western art and philosophy, Frobenius poses a return to unity as his methodological goal. Like Spengler, he proposes the return of science and philosophy to myth and poetry:

> Another lesson that we may learn from African studies is that of the organic character of primitive cultures, which unite tectonic with monumental features, ideas with practical effect and facts with a conceptual content. As a result we receive flashes of enlightenment which continue to inspire the human mind and will obliterate the differences of method that are still customary in approaching "primitive" and "historic" cultures respectively. This means that it will be possible to make a scientific study *of the whole of human culture* from its beginnings, *as an organic unity.* (Frob, 54–55)

Here the translation makes Frobenius's goal ambiguous: is it a unified "culture" or a unified "science"? It is probably both, since Frobenius inscribes a hermeneutical circle, as opposed to any number of the less traditional and metaphysical figures which Pound uses to graph Kulchur, even if his own version of paideuma is palimpsestic and simply "jagged" or, like Malatesta's "post-bag," irrecoverable.

This more traditional or teleological Frobenius is hardly hinted at in Pound's text. When he credits the "archeologist" with system or new method, it is one that conforms to his own Kulchur, a collection of suggestive fragments and the proliferation of methods, not one method. He acknowledges Frobenius's special (notion of) intuition: "He has in especial seen and marked out a kind of knowing, the difference between knowledge that has to be acquired by particular effort and knowing that is in people, 'in the air'" (GK, 57). Here Pound might be thought subtly to adapt, or to put into active circulation, Frobenius's more rigorous "intuition," as he takes yet another precursor further outside the orbit of systematic philosophy and into the general culture from which he claims to draw his own method and insights: "When I said I wanted a new civilization, I think I cd. have used Frobenius' term" (GK, 58). Thus, Pound's first use of Frobenius against the literary-philosophical establishment or, more exactly, against the old forms of cultural history and the mechanical and abstract thought of Germanic philosophy, is to read Frobenius against himself, turning his dream of a unified system into one among many fragments. In the following, Frobenius is made to endorse detail against generalization, and open interpretations ("whatever my generation can offer") in place of method, or so Pound has instructed us to read "the sign of the mortar":

> The "New Learning" under the ideogram of the mortar can imply whatever men of my generation can offer our successors as a means to the new comprehension.
> A vast mass of school learning is DEAD. It is as deadly as corpse infection.
> CH'ING MING, a new Paideuma will start with that injunction as has every conscious renovation in learning. (GK, 58)[33]

Here, Pound might be thought to hold in suspension, but not to unite in one method, the two master tropes of the *Guide,* the mortar which grinds knowledge and the paideuma, the play of languages and cultures which transforms learning into action and facts, into (electrical or rhetorical) power. Frobenius, who with Spengler hoped to modify the Western, Classical prejudices, becomes but one point of reference. This is not to say, however, that Pound failed to grasp the implications of Frobenius's work for the philosophical tradition.

He does not adopt wholesale the eschatological (and entropic) elements of Frobenius's *Kulturmorphologie.* He chooses instead to interweave some of its more vexed details into his own notion of Kulchur, taking advantage of the slippages in Frobenius's terminology for his claim that Kulchur is irreducibly phenomenal, a bundle of sensible fragments and not a spiritual unity. Notwithstanding the insistence on "proper naming," "clear definition," and "the *mot juste* (perhaps Pound's recurrent piling up of words suggests a certain irony and not simple indecision or haste) which opens the *Guide,* he uses "paideuma" in a number of contradictory senses, until it virtually marks the translative moments by which Pound hopes to renovate culture and cultural history (to "make it new"). In this way, paideuma can mean anything from art's special truth to a good meal. For example, in a sense that surpasses Eliot's circular formalism, he suggests that paideuma is each artist's original intuition which can be read in his work: "How to see works of art? Think what the creator must perforce have felt and known before he got round to creating them. The concentration of his own private paideuma" (GK, 114). In contrast to this metaphysical exchange of Truth, paideuma is also the national habits of modern Europe represented by restaurant menus, or simply the variety of good meals available to the "cultured" traveler: "Le Voyage Gastronomique is a French paideuma. Outside it you can get English roast beef in Italy (if you spend 25 years learning how), you can find a filetto of turkey Bolognese" (GK, 112).

In Pound's text, "paideuma" can name almost any detail of everyday life, art, science, philosophy, or religion; that is, all those things that he also calls Kulchur. Its appearance signals passages where he, perhaps, through a rather perverse "intuition," gathers together those details especially anathema to traditional cultural history. He brings this double register, which marks both the pervasiveness of writer's and reader's cultural assumptions and the figurative possibilities of his borrowed vocabulary, to bear on his readings of Aristotle and the pre-Socratics. In this way, he measures his personal paideuma against Aristotle's "Greek paideuma," and, at the same time, he shows that poetry's role in setting ideas into action far exceeds that of philosophy's, that poets and not scholars should be the keepers of Kulchur. In a sense discussed earlier, paideuma is a catachresis that stands for all catachreses, an idea with two names that show(s) that ideas resist naming.

IV

Pound's argument with a certain residual Aristotelianism—or the general philosophical tendencies toward abstraction and categorical judgments which he traced back to the Renaissance adoption of Aristotle into the philosophical canon—runs through *Guide to Kulchur,* only to surface in a protracted reading of *The Nicomachean Ethics* near its conclusion. This scattering of fragments is strategic: not only does it prove Pound's claim that his intermittent and residual Kulchur is more active—and activating—than the traditional (notion of) culture, but it is also the best way to disclose—or dissociate—those Aristotelian assumptions which have become commonplace. Pound's multifaceted examinations unveil Aristotle's lapidary position and his narrow definition of the philosopher's role. Thus, in an early section, "As Background," he traces the need for the New Learning, or active thoughts and responsible behavior, back to the Greek origins of philosophy:

> Aristotle was so good at his job that he anchored human thought for 2000 years. What he didn't define clearly remained a muddle for the rest of the race for centuries following. But he did not engender a sense of social responsibility. (GK, 39)

Pound focuses, then, not on Aristotle's precise definitions, but on his limited view of philosophy, that is, his refusal to consider the ethical alongside the aesthetic and the practical. Aristotle, he suggests, is responsible for the rigid definition of culture he must attack throughout the *Guide.* Later, in the close reading of the *Ethics,* perhaps invoking Frobenius's opposition of mechanical to organic thought, he confirms this opinion of Aristotle: "Master of those that cut apart, dissect and divide. Competent precursor of the card-index. But without the organic sense. I say this in the face of Aristotle's repeated emphases on experience, and of testing by life" (GK, 343).

Pound argues that, contrary to the popular wisdom and the history of philosophy, Aristotle lacked any sense of culture as a whole. In fact, that original—though by Pound's account virtually the latest—systematic philosopher was instrumental in separating out poetry—and poetic or figurative language—from the scientific and philosophical discourses. With the attempt to systematize is born the schismatics that precludes System. Pound's recurrent, interested objection to Classical ethical literature is that it assigned poetry and art subsidiary roles in education and government. He detects this schism in Aristotle's quibble on *theoria* and *eudaimonia.* But he also notes the wider repercussions of clear divisions of knowledge into aesthetics, ethics, politics, logic, and the like. Citing a modern textbook on the history of philosophy, perhaps because it is closer to general Kulchur than any academic interpretation, he concludes that an abandonment of a "totalitarian" discourse coincides with philosophy's abrogation of its pedagogic responsibility:

> This next circuit showed a frittering away of the totalitarian concepts: "Ethics and politics are no longer one." . . . *Knowing is no longer the basic problem "of philosophy,"* "religious tendency" and concern with practical end sprout, and prevail over curiosity. (GK, 120; emphasis added)

For this reason, Aristotle represents a falling away from a more complete philosophy, perhaps from the Confucian New Learning Pound tries to inculcate. Pound argues that Aristotle had lost an original vigor and curiosity, if not the sort of intuition Frobenius associated with a primordial unity, and Gourmont with a "spurt" of seminal thought.

Pound does not try to heal the breach between philosophy and poetry by unifying the two discourses or proposing, like the late, nostalgic Heidegger, a return to a pre-Aristotelian—or even the pre-Socratic—fullness. Instead, elliptically and often without comment, he juxtaposes fragments of poetry and philosophy. Such strategies as positioning Dada and Gorgias in attacks on Aristotelian abstraction suggest that poetry and rhetoric have always disturbed and fecundated philosophy, and vice versa.

Pound uses such citations to show that even the earliest poetry has a
critical function; that is, it questions, rather than grounds, the version of
the tradition which subscribes to a truth prior to language. Quite apart
from other (Romantic and Vician) readings of the origin of abstract philosophy out of an even more abstract poetic unity, he assigns Homer the
task of correcting Aristotle:

> You cannot get anything DONE on an amoral tradition. It will merely slide
> down. Arry was interested in mind, not in morals. It was in the air, it was
> there, the decline of the hellenic paideuma. This Homeric vigour was gone,
> with its sympathetic rascality, its irascible goguenard pantheon. The splin-
> tering . . . had begun. (GK, 331)

Here the decline of the "hellenic paideuma," the whole series of things
and "inrooted ideas" which make up a culture, into the metaphysical "mind"
indicates "the slide down" from poetry to philosophy. Yet Pound's version, as opposed to the reigning mythopoetic and/or philosophical one,
is a "fall" from plurality into unity, from figurative into abstract language. It is altogether appropriate that this parody should turn on a particularly disruptive or modernist poetic allusion and a translation. Thus,
in the position of logos and *Logos,* the Word as well as the singular God
to which Aristotle and St. John trace the origin, Pound places Homer's
"goguenard pantheon," recalling at once that the Greek poet drew inspiration from more than one god and that the Trojan Wars were instigated
by an Olympian domestic quarrel. *Goguenard,* a French word that can
be translated as "bantering," "jeering," or "mocking," suggests, then, a
certain Homeric playfulness against Plato's more sober *paideia.* But why
the French? One cannot be certain whether he intended an allusion to
Cocteau, the playwright who was, by Pound's estimation, both Dadaist
and Classicist. Yet earlier in the *Guide* he praises the Frenchman's "translations" of Greek tragedies onto the modern stage—though most would
hesitate to call *Oedipus,* his multimedia avant-garde extravaganza, simply
a play, let alone a translation. In this context—and by a pun on "play,"
as "drama" and as (Homeric?) "mischievousness"—Pound revises the
traditional narrative of the fall away from Homer:

> *There was magnificence; there was a SENSE of play.* . . . Nobody but a
> fanatic like myself wd. have the crust to insist that *greek writers were on
> the down grade AFTER Homer.* . . .
> Yet the greek drama exists. Cocteau by sheer genius has resurrected it.
> Antigone "T'as invente la justice" rings out in the Paris play house *with
> all the force any man ever imagined inherent in greek originals.* (GK, 93;
> emphasis added)[34]

In a somewhat more serious vein, Pound traces the parallel descent of Greek philosophy from Heraclitus, suggesting that the language of the Stoics had already undergone the loss of poetic force and was, in Frobenius's word for "modern" thought, "mechanical":

> Even the Stoics were lured "out on the limb," they end up talking about ekypyrosis, about which they knew no more than you, me or Heraclitus.
> "From god the creative fire, went forth spermatic *logoi,* which are a gradual and organic distribution of an unique and spermatic word (logos)."
> That's all very nice, and it even contains a provision against the process being merely mechanistic, but like all other generalized statements it tends *toward* mechanistic. (GK, 128)

Pound alludes to Zeno's elaboration of the Heraclitan fragments into a systematic cosmology, indeed, a mechanistic "uni-verse," grounded, as it happens, in Aristotle's *Logic.* While for Pound the "provision" against the mechanistic reproduces the mechanistic, which is to say, metaphysical abstraction, the general thrust of the idea which Zeno derives from Heraclitus, that in the beginning there was spermatic force and multiple gods, or *logoi,* rather than a univocal word, or *logos,* is a notion Pound found attractive from the time of his earliest criticism.[35] That these gods are associated with the "word" indicates Pound's identification of the ancient gods with a specifically rhetorical or figurative linguistic force.

If this appears to be a most unorthodox Aristotle, one can explain it by noting that Pound's strategy in the *Guide,* as in his earlier polemical writing, has been to turn the classical texts of scholastic philosophy against themselves in order to open out their "poetic" nature. Late in the *Guide,* he describes the broad outlines of such reading, reflecting, at the same time, on the (dis-)organization of his book:

> I have left my loose phrases re Aristotle in Section I, *as sort of gauge,* the reader can see for himself what *my residuum of opinion was before examining the Ethics,* and measure that by the notes set down during my examination of that particular text. (GK, 347; emphasis added)

One would be naive to take Pound's words at face value. This is to say that he might very well have cheated by reading Aristotle before writing his "loose phrases," and carefully enough to have incorporated some of the Greek philosopher's figures willy-nilly into various fragmentary poems and essays. Nevertheless, comparisons of the sort Pound calls for do not yield the proof, let alone the clear lines of influence, demanded by literary historians. The measurement of "residual opinions" by "notes" *ends* not in a final or authoritative interpretation, but in a reading process that involves the multiplication of readings.

Pound insists upon this openness and upon figures in lieu of system, when, drawing upon the synecdochal model of paideuma, he identifies *The Nicomachean Ethics* as " 'the most Aristotlean' " of the philosophical texts, and its writer as an "indication of where the Western mind or one western mind had got to by, say B.C. 330" (GK, 344). The claim that Aristotle represents Greek philosophy is hardly startling, but Pound complicates this by treating him not so much as the writer of a particular book but as a reader of earlier Greek thought or, better, as a receptacle of a certain Homeric and Heraclitan residue: "I am not attacking the conscious part of Aristotle, but the unconscious, the 'everyone says' or 'everyone admits' " (GK, 331). When he turns to specific passages, or to etymologies of individual words like *theoria* and *eudaimonia,* he finds that Aristotle has thinned or tamed the ideas of the earlier Greek thinkers which his own—more "poetic"—readings have put back into circulation. For example, "I get the feeling that Arry is like a shallow, clear layer of water, now and again flowing over the deep that is, the thought of more compact and fibrous precursors" (GK, 336). In a passage we have already considered, he calls upon Socrates as well as Heraclitus to make explicit his complaint that Aristotle's abstract or philosophical language, indeed, his "logic-chopping," marks a decadence in the abandonment of figurative or poetic language for philosophy:

> The *Socrates* of the Dialogues *used to do this kind of thing with more imagery.*
> The profound flippancy of a decadence, when men don't know when to stop talking and have ceased to respect the unknown! Very well! *Arry teases Heraclitus for vehement assertion, but is that assertion any worse than silly logic-chopping* on the same matter? (GK, 340; emphasis added)[36]

That concludes a forty-page "close-reading" of the *Ethics,* in which Pound has subjected Aristotelian logic to a variety of rhetorical dissociations, private allusions, loose analogies, and name-calling. Aristotle becomes, for example, a modern journalist ("were he alive today, wd. be writing crap for 'Utilities' ") and the source of Mussolini's (however modest) errors: "The things still needing to be remedied in the Italian state are due to an Aristotlean residuum left in Mussolini's own mind" (GK, 306, 309). Moreover, citing the translator of his Loeb edition, he says that Aristotle's book is really just lecture notes, probably hastily jotted down by Nicomachus, Aristotle's son. This fracturing and reassembling of Aristotle's discourse within a contemporary rhetoric brings Pound to claim that the *Ethics* is "heteroclite, a hodge-podge of astute comment and utter bosh, material for a sottisier" (GK, 308), that is, an example of the sort of satiric blurb Pound insisted should be part of every newspaper and literary review. Thus, Pound's destructuring, if not deconstructive, reading mimes in an approximately discursive form what *The Cantos* have been doing in a more or less poetic form. And in a fragment titled "Incipit B" Pound seems to announce the mutual undoing of the critical and poetic texts from which his own text is hardly immune:

> PROSE in the main presents things with full or fullish explanations.
>
> POETRY represents them rather in the sudden or at least unexplained and unexplaining way wherein they strike the mind through the emotions or stir one via the senses, or even blur and confuse the perceptions or render them fragmentary chaotically vivid.[37]

At this point, by way of halting another potentially endless reading back and forth across the fine lines separating *The Cantos* from Pound's late critical prose, we might note that he had earlier used "heteroclite" to describe his own *Guide,* if not the bulk of his prose:

> I am, I trust patently, in this book doing something different from what I attempted in *How to Read* or in *ABC of Reading.* There I was avowedly trying to establish a series or set of measures, standards, voltometers, here I am dealing with a heteroclite series of impressions. (GK, 208)

"Heteroclite," not an everyday word, means deviating from common forms or rules, specifically grammatical rules. The noun "heteroclite" refers to one who writes ungrammatically as well as to ungrammatical noun declensions. Thus, it would seem to describe Pound's own fragmentary notes, but hardly the books of the founder of logic and the philosophical categories. And Pound's self-reflection subverts its own claim to any sort of mastery, recalling, in fact, Nietzsche's observation that reader and writer are grammatical fictions.

Michel Foucault or, more precisely, his translator, offers a definition of "heteroclite" that indicates the critical thrust of Pound's text, which goes far beyond the coincidence of one word:

> There is a worse disorder than that of the *incongruous,* the linking together of things that are inappropriate; I mean the disorder in which fragments of a large number of possible orders glitter separately in the dimension without law or geometry, of the *heteroclite;* and that word should be taken in its most literal, etymological sense: in such a state, things are "laid," "placed," "arranged" in sites so different from one another that it is impossible to find a place of residence, to define a *common locus* beneath them all.[38]

Foucault describes the "dread," the mixture of fear and amusement, he felt when he imagined an "archeology of the human sciences" that might accommodate Borges's library. One might well argue that Pound's *Guide* is a modernist analogue of the Borgesian library, though it sustains little of the irony and less of the despair than Foucault discovers in that uncanny (*unheimlich*) "place of residence."

Pound's *Guide,* then, is an anomaly even among his own texts. Rather than a cultural compendium or even a *Tempio*-like structure of cultural fragments, it is in a sense a "canto" or even, as etymology and homology permit, a provisional "room" ("Rooms, against chronicles") housing the signs out of which Pound hopes to fashion a future "city," his metaphor in the later *Cantos* of a successful culture or earthly paradise—even if, in Pound's view, all such cities, from Ecbatana to Dioce, are subject to a rhythm of growth and decay. To superposition the *Guide* and *The Cantos,* to argue that the former is a kind of index to *The Cantos*—and, strangely enough, an index that is at the same time prospective because it not only records the details of completed work but is a storehouse for future constructions—is by no means arbitrary. Pound himself, in several passages, linked his paideumic text to his open and ongoing poem. For example, in an unpublished essay, curiously entitled "Notes Toward a Preface," which Pound once thought of as introducing his collected prose, he virtually calls the *Guide* a preface to the long poem. This kind of preface, we have come to see, destroys prefaces, since it cannot anticipate any future development that completes a total structure:

> The reader may prepare for notes on a preliminary survey, the end product
> of which, or one of the end products so far as the writer is concerned,
> being a poem of some length (now resting at its 40th Canto).[39]

One of the effects of such prefatory games, or "foreplay" in the Derridian sense, is to violate the strict demarcations between prose and poetry, as between literature and philosophy or criticism, producing what Pound had anticipated from the beginning, an interpretative poem.[40]

Paul de Man, commenting on the endless self-reflections in the Nietzschean text, describes this sort of intervention as that "rhetorical mode" by which the literary and critical discourses are simultaneously created and destroyed: "Philosophy turns out to be endless reflection on its own destruction at the hands of literature."[41] The thrust of Pound's writing adds one more turn of the screw, for the consequences of the death of philosophy imply the death of literature in the classical sense—that is, the notion of the hermetic or closed poem—and thus of culture in the classical sense—that is, of culture as defined in this or that book.

In a strangely prophetic passage early in the *Guide,* Pound defines his new learning as a different kind of cultural text, not referential to the past but prospective, or even, as Charles Olson would later define it, "projective."[42] He claims that he is writing his paideuma not out of a library but on the model of a new kind of memory:

> In the main, I am to write this new Vade Mecum without opening other volumes, I am to put down so far as possible only what has resisted the erosion of time and forgetfulness. And to this there is material stringency. Any other course wd. mean that I shd. quite definitely have to quote whole slabs and columns of histories and works of reference. (GK, 33)

A few years later, history would provide the ironic occasion of just such a scene of writing. Though Pound obviously did not construct the *Guide* without opening other volumes, his way of reading does suggest, as we have seen, an active dislocation of reference, a kind of broken reading. That is to say, the *Guide* indexes *The Cantos* which are in themselves an index of cultural memory. But by the end of World War II, imprisoned at Pisa, Pound discovered himself not simply in a prison house of language but in a literal cage, with a minimum of reference or organizing texts to assist him in ordering the chaos without as well as the despair within. Canto LXXIV is, in its way, a rewriting of the *Guide*, undertaken in order to reassemble the materials of a poem that lay scattered in Pound's memory and in a modernist canon dominated by Eliot. At Pisa, he begins his poem all over again as a compendium of modern Kulchur and ancient beauty, an affirmation as well as a propadeutic of ways of ordering these fragments.

V

Without pretending to offer an exhaustive reading of this Canto, which is as "open"—indeed, as "heteroclite"—as any of Pound's texts, I would like provisionally to end by considering how he rejects the two courses presented by Eliotic or "classical" modernism: despair and alienation on the one hand, religious or mystical conversion on the other. In the face of these tempting alternatives, which represent opposite ways of refusing history and change, Pound recalls his dream of a city built on the model of *The Tempio,* that is, a palimpsestic layering of textual and architectural plies or figures recording historical and cultural transformations. In contrast to such indecipherable locales as the "Unreal City" of *The Waste Land* (from Jerusalem to London, or from the eternal and mystical biblical city to its secular repetition), Pound's metamorphic city gathers his previous allusions into a series of ideal cities. These cities, or metaphors of metaphor, serve as repositories for his own future interpretive and artistic creations, his continuation of an interrupted interpretative poem. His cities, and those temples (sacred) or museums (secular) on which they appear to be centered, become figures of an archival, nearly meta-poetic language to which every poet has recourse. Thus, Thomas Jefferson can adopt the inscription on Isotta's sepulchre, "Tempus tacendi Tempus loquendi," or, "There is a time to be silent, a time to speak" (LXXIV: 429, *et passim* [Ecclesiastes 3:7]). This transposition of Italy into America, a set of ruins into a ruined dream, marks at once the ravages of time on Pound's memory and the temporization or delay his meditative—silent or reflective—poem "shores" against the horror of his circumstances which nevertheless remain inscribed within his very poem: "beyond the stockade there is chaos and nothingness" (LXXX: 501).

To the few texts available to him, the Confucian *Great Digest,* the U.S. Army *Maintenance Manual,* the Bible, and a discarded *Pocket Book of Verse* recovered from the outhouse, he brought the themes of the earlier Cantos and the procedures of the *Guide.* Most important among the latter are his piling up of "luminous" or "ideogrammic" detail and his use of transformative or "paideumatic" figures. While his ideas and tropes cannot be reduced to one character or mastered by a central metaphor, his declaration to rebuild his city in the form of "temples *plural*" (emphasis added) and in the names of the pagan deities, if not the "*hilaritas*" (LXXIV: 444) of the Homeric pantheon, is repeated at crucial points in the Canto.[43] Here, for example, his fragments abbreviate or index the (United States?) constitution, the legendary and biblical city of Ecbatan (sometimes "Ekbatana") built by the Babylonian King Dioce, and Terracina, the modern Italian city where Pound hopes to replace—even to reclaim from her natal sea—the statue of Venus in the ruins of the pagan temple. All these monuments and figural repetitions are subject to historical or temporal process and subordinated to another of this Canto's recurrent tropes, the metamorphic force or "wind" ("Zephyr" and elsewhere "scirocco") and "water":

> I surrender neither the empire nor the *temples plural*
> nor the constitution nor yet the city of Dioce
> *each one in his god's name*
> as by Terracina rose from the sea Zephyr behind her
>   and from her manner of walking
>   as had Anchises
>   till the shrine be again white with marble
>   till the stone eyes look again seaward
>   *The wind is part of the process*
>   *The rain is part of the process*
>   (LXXIV: 434–35; emphasis added)

Pound overlays, or superpositions, a biblical reference with "pagan" themes. By quoting or incorporating a fragment (even more importantly, by *not* quoting the whole sentence which would radically alter the thrust of his own lines), he invokes "gods" in the place of the Judeo-Christian God. The biblical verse, Micah 4:5, reads: "For all the people walk each in the name of its god, but we will walk in the name of the Lord our God for ever and ever." Here, Pound's selective quotation, or abusive reading of a text and thus of a whole tradition, might be thought to recall his citation of the Heraclitan logoi against the metaphysical logos or Christian Logos. And, in passing, we might note that he repeats and revises this phrase throughout the Pisan sequence until it intersects with a certain Dadaist undercurrent: "each in the name of its god" (LXXIV: 441); and finally "Teofile's bricabrac Cocteau's bricabrac/seadrift snowin' 'em under/everyman to his junk-shop" (LXXIV: 453).

Moreover, Ecbatan refers us back to earlier Cantos, especially to Canto V, where it is both city and reference book and already enfolded by layers of Classical and biblical references:

> Great bulk, huge mass, thesaurus;
>   Ecbatan, the clock ticks and fades out
>   The bride awaiting the god's touch; Ecbatan,
>   City of patterned streets; again the vision:
> Down in the viae stradae, toga'd the crowd, and arm'd
> Rushing on populous business
>
> (V: 17)

Ecbatan appears in Herodotus's *History* and in legend, but let us pursue its biblical traces. Perhaps because Pound repeatedly affiliates himself with such figures as Homer's Elpenor, " '*A man of no fortune and with a name to come*' " (I: 4; see also LXXIV: 439), and with Wanjina or Ouan Jin, the mythical figure who created too many things by talking, " 'I am noman, my name is noman' " (LXXIV: 426), the story of his namesake does not seem to be recorded in his text. Nevertheless, Ezra, Pound's Christian name, recalls a minor prophet instrumental in rebuilding the Temple at Jerusalem. At one point in the Old Testament narrative, Ezra is sent by Darius, the Median king, to search for the Decree of Cyrus in the archives of the old Babylonian city of Ecbatan. I will cite a verse particularly full of the kind of sacerdotal imagery associated with the Ecbatan of *The Cantos:* "In Ecbatan, the capital which is in the province of Media, a scroll was found in which this was written. 'A record. In the first year of Cyrus the King Cyrus issued a decree: Concerning the house of God at Jerusalem, let the house be rebuilt where sacrifices are offered

and burnt offerings are brought'" (Ezra VI: 2–3). If Pound alludes, even
in the most indirect way, to the biblical text, it is not to the prophetic
utterance or voice but to the "thesaurus" or treasure house of language
of Ecbatan, a "decree" inside a "record" in the archives at the center of
the "patterned city." We are now back inside the heteroclite "library"
which at every stage signifies the absence of a point of reference, locus,
or fixed center. Ecbatan, then, is not only repeated throughout *The Cantos;* it has an aberrant text inscribed at its center. Where one might expect
to find the divine Word or at least the stable patterning of the Babylonian
streets, there is instead a vault of texts, a crypt not unlike *The Tempio*
housing its own arche-texts and holding the remains of the architects. But
let us halt such speculations, such specular plays of imagery hinging as
they do on accidents of meaning and the indeterminacy of the poet's intention, and return to more immediate contexts of Canto LXXIV.

In addition to indexing the earlier *Cantos* and at least one ancient tradition, the fragments in the passage cited above subtly "read" the dominant imagery of *The Waste Land* and the (Christian) resignation of "The
Hollow Men." In Pound's poem, water is "part of the process," the metamorphic element which undoes and reestablishes the temples and the patterns of *The Cantos;* in *The Waste Land*'s "Death by Water," it signals
the eschatological movement that can only be effected by divine intervention. If the allusion to Eliot, and the replacement of his monotheism
by Pound's own barely named pantheon, is buried here, one might return
to the beginning of Canto LXXIV. Picking up the thread of an argument
with Eliot which runs throughout *The Cantos,* surfacing in, for example,
a battle between the muse of history and truth itself over the status of
textual fragments—"These fragments you have shelved (shored). /'Slut!'
'Bitch!' Truth and Calliope/Slanging each other sous les lauriers" (VII:
28)—Canto LXXIV opens with a reference to the nursery rhyme which
ends "The Hollow Men," and in a context which parodies the Crucifixion:

> The enormous tragedy of the dream in the peasant's bent
>   shoulders
> Manes! Manes was tanned and stuffed,
> Thus Ben and la Clara *a Milano*
>   by the heels at Milano
> That maggots shd/eat the dead bullock
> DIOGONOS, Δίγονος, but the twice crucified
>   *where in history will you find it?*
> *yet say this to the Possum: a bang, not a whimper,*
> with a bang not with a whimper,
> To build the city of Dioce terraces are the colours of star
>   (LXXIV: 425; emphasis added)

"Possum," the animal who can pretend to be dead while alive, is Pound's nickname for Eliot. Pound evokes the spirit of his friend in order to revise the line, "Not with a bang but a whimper," and to suggest not an apocalyptic or entropic end to the world but an explosion ("blast") of images and the proliferation of gods. Out of the grotesque hanging of Mussolini and his mistress, Claretta Petacci, Pound generates a pantheon and several of the themes and tropes of his earlier writing. "Manes," for example, names both the Roman spirits of the dead and the founder of Manicheanism, the Gnostic heresy that informed the mystical verse of Cavalcanti and Pound's other favorite Provençal writers. "Clara," of course, means light, perhaps the *same* transformative light the Manicheans considered prior to and more powerful than the Judeo-Christian God. And Manes, flayed alive for his beliefs, is a "subject rhyme" as well as an echo of Mussolini. The appearance of Dionysus, to whom Pound had earlier referred by way of Ovid's *Metamorphoses* (most strikingly in Canto II), signals transformation and the survival of an ecstatic and intoxicating poetic force beyond the destruction of various forms. Further, "Diogonos" (Pound's spelling is usually "Digones"), literally "twice-born," recalls the tangle of myths concerning the parentage and birth of the Greek god and suggests the originary figuration and multiplicity Pound associates with natural or primordial language.

This process of reading multiple allusions out of a historical event, of equating not only Mussolini but also a series of pagan gods with the Crucifixion, at once undoes history and the Christian notions of the eschatalogical Christ and his redemption of the individual. Thus, Mussolini will not be reborn as Mussolini, because he is only a figure for Pound's own interrupted vision. Unlike *The Waste Land,* where Frazer's "hanged-god" and the Fisher King are analogues and imperfect prefigurations of Christ, *The Cantos* do not offer one God to which all myths can be made to refer or one metaphor which orders all the other tropes. Pound proposes neither an alternative religion nor an escape from history into mysticism. Instead, by reversing the essentially en-tropic Christian teleology, he resumes his own poetic construction of Ecbatan, that poetic structure which counts Dioce, Malatesta, Frobenius, Mussolini, and Pound among its "live-in"—or is it "encrypted"?—builders.

Later, in Canto LXXIV, Pound recalls details of *The Waste Land*'s urban landscape and its key metaphors of the Thames and death by water. More than once, autobiographical and historical fragments surface as though to modify Eliot's bleak account of modern detritus. Pound's London and Paris, while every bit as chaotic and cacophonous as Eliot's, are inhabited, if not by muses and once "departed nymphs," by his favorite artists and *bricoleurs.* Thus his rather whimsical elegy of those "Men of '14"

who engaged the whole modern paideuma, writers now recalled as metonymic fragments out of which Pound is weaving his own song:

> Fordie that wrote of giants
> and William who dreamed of nobility
> and Jim the comedian singing:
> "Blarney castle me darlin'
> you're nothing now but a StOWne"
> and Plarr talking of mathematics
>
> (LXXIV: 432–33)

They are, these writers, now reduced to their texts, which endure, as
Pound notes of Joyce's "voice," like an emblem of a structure once thought
complete and invulnerable. They are now signs of a "process" Pound is
trying to re-member as the literary tradition—as he relimbed (or "Gathered the Limbs of") Osiris, in the fragmented essays assembled under
that title.

Allen Upward, the mystic and amateur sinologist who was one of Pound's
earliest London associates, makes a brief appearance in a passage which
evokes *The Waste Land* and Pound's rather different notion of the transformative properties of water:

> sd/old Upward:
> "not the priest but the victim"
> his seal Sitalkas, sd/the old combattant: "victim,
> *withstood them by Thames* and by Niger with pistol by Niger
> *with a printing press* by the Thomas bank"
> *until I end my song*
> and shot himself
>
> (LXXIV: 437; emphasis added)

The echo of the refrain Eliot had borrowed from Spenser's *Prothalamion*
("Sweet Thames, run softly *till I end my song*") might be thought to lend
a bitter irony to his old friend's death. Just so, in 1936, Pound had attributed Upward's suicide to despair over his own bugbears, the absence
of a reliable scale of literary values, and the choice of Nobel laureates
in the field of literature: "Have always thought poor old Upward shot
himself in discouragement on reading of award to Shaw. Feeling of utter
hopelessness in struggle for values" (L, 284). The tracing of allusions to
Upward cannot end there, for he is quoted in *Guide to Kulchur* and again
associated with water: "'The quality of the sage is like water,' I don't
know the source of Allan [sic] Upward's quotation." To this fragment
Pound appended a cryptic note which bears on his readings and on our
own: "This text is interpreted in various manners" (GK, 84). Thus, by
tracing another line of Pound's concatenating self-reflections, we arrive

at a textual abyss where another, entirely different, Nietzschean or affirmative irony reigns.

Pound's acceptance of his own scriptive, if not his historical, "Fate," his version of the "philosopher's laugh" at the prospect ("pro-spect") of infinite interpretation can be detected in the closing lines of Canto LXXIV, where water is again part of the process which at once creates and destroys order:

> This liquid is certainly a
>   property of the mind
> nec accidens est but an element
>   in the mind's make-up
> est agens and functions dust to a fountain pan otherwise
>   Hast 'ou seen the rose in the steel dust
>   (or swansdown ever?)
> so light is the urging, so ordered the dark petals of iron
> we who have passed over Lethe.
>
>   (LXXIV: 449)

Surely those critics who have not without nostalgia cited this instance of Pound's recurrent metaphor ("the rose in the steel dust") as proof of his belief in a recoverable meaning behind the "verbal manifestations" of his own texts and a "patterned energy" to the chaos of historical detail, are not wrong. Nevertheless, this series of heterogeneous quotations and formulas is disordered from within. A dissemination of images, which is here figured as the "liquid mind" and elsewhere as "the spermatic brain," undoes his own oppressive patterning. The "dark petals of iron," a reading of "the rose in the steel dust," almost literally recalls the literal prison cage and Pound's own entrance into the underworld which opens *The Cantos* with the admonition of Tiresius to return to bury Elpenor, he of "no fortune" and with a "name to come." What is at work here as the process is at once a wearing away or washing away of the received ideas and structures of the tradition and a regathering (a favorite Pound figure) of the fragments or *metonyms* (a trope which can involve the violent synecdoche of condensing a writer's corpus into his name) into the moving patterns of a new yet prospective structure.

Canto LXXIV is structured by a number of Pound's own metaphors for the poetic process, metapoetic figures, as it were, which are also made up of still earlier textual fragments: the "paideuma," the "periplum," the "rose in the steel dust," as we have noted, along with the ideogram Ch'ing Ming and, of course, the various cities that at once repeat and transform earlier cities or, like the culture of Waguda and Ecbatan, rebuilt again and again, "ply over ply." No figure by itself is sufficient to "the process," though each is a figure of process and transformation—that is, each is a trope of trope, each repeating yet opening the others, like the explicative "petals" which at once represent ("dark petals of iron") and break ("the rose") the fixed pattern of the cage. Turning on each other, these interpretive metaphors enable Pound to resume the writing of his poem, and they transmit that energy, both ideogrammatic and paideumatic, which "pulses through all things."

Canto LXXIV, therefore, begins again a poem that is inaugurated in a repetition, or a return; that is, in a translation, both literally and figuratively, of Homer's "sightless narration" (Canto VII). By appropriating the master figures of others into a new library of metapoetic tropes, Pound does what Nietzsche, in "Truth and Falsity in the Extra-Moral Sense," says is the "work" or the "dreamwork" of all art and the very "impulse" to metaphor:

> The impulse toward the formation of metaphors, that fundamental impulse of man, which we cannot reason away for one moment . . . is in truth not defeated nor even subdued by the fact that out of its evaporated products, the ideas, a regular and rigid impulse seeks for itself a new realm of action and another river-bed, and finds it in *Mythos* and more generally in *Art*. This impulse constantly confuses the rubrics and cells of the ideas by putting up new figures of speech, metaphors, metonymies; it constantly shows its passionate longing for shaping the existing world of waking man as motley, irregular, inconsequently incoherent, attractive, and eternally new as the world of dreams is. For indeed, waking man per se is only clear about his being awake through the rigid and orderly woof of ideas, and it is for this very reason that he sometimes comes to believe that he was dreaming when that woof of ideas has for a moment been torn by Art. (TF, 513)

POSTSCRIPT

Rhizomatic America

Far too as her splendors shine, system on system shooting like rays, upward, downward, without center, without circumference—in the mass and in the particle, Nature hastens to render account of herself to the mind. Classification begins . . . and so, tyrannized over by its own unifying instinct, [the mind] goes on tying things together, diminishing anomalies, discovering roots running under ground whereby contrary and remote things cohere and flower out from one stem.

EMERSON, "The American Scholar"

A PACT
I make a pact with you, Walt Whitman—
I have detested you long enough.
I come to you as a grown child
Who has had a pig-headed father;
It was you that broke the new wood,
Now is a time for carving.
We have one sap and one root—
Let there be commerce between us.

POUND, *Personae*

At times, Pound's desire for a compatible and legitimate tradition led him to renounce his Americanism and to abjure modernism altogether. Nevertheless, whether under the name of "Make It New" or "The American Risorgimento," his historical and/or aesthetic project is articulated in terms which belie any simple desire to keep the tradition intact. He wanted to *drill* a certain version of America into modern letters, to *graft* fragments of modern life onto the trunk of European art, and at the same time to carve out an American heritage. I emphasize Pound's recurrent metaphors of grafting and drilling, of the organic and even the sculptural as against the architectonic. These figures, endemic to American poetics, at once ground and unsettle the edifice—or is it the system of poetic roots and shoots?—Pound hoped to erect or resurrect as an American tradition. One

211

might note in passing that an American tradition, let alone one that combines the mutually exclusive models of the organic and the monumental, has always seemed oxymoronic if not downright impossible.

In "What I Feel About Walt Whitman," a note written in 1909 but left unpublished until 1955, Pound outlined a strange genealogical reconstruction in characteristically mixed metaphors and catachreses.[1] Presenting the old saw, that is, "the family tree," perhaps too graphically, he says of Whitman that "the vital part of my message, taken from the sap and fibre of America, is the same as his" (SP, 145). This American tree, of which Whitman and Pound are parts, though not simply living trunk and branch, is indeed an unnatural one. Growing backward in time, it has not yet taken root. In fact, Pound suggests that both he and Whitman are misplaced and untimely. If Whitman was an imperfect beginning, Pound is the repetition of such a beginning. In this way, Pound marks both a repetition of and a revolution against Whitman, who, for better or worse, found himself in a similar position of at once taking up and overthrowing his European and American heritage(s).

Pound and Whitman are hardly alone in recognizing America's ambiguous cultural imperatives. Indeed, Henry Adams, for whom American culture was inescapably problematic, translated the search for roots into the broader and no less heterogeneous areas of pedagogy, Medieval art history, and the new sciences of geology, genetics, and physics. Adams recognized that Americans' need for cultural legitimacy necessarily involved the breaking of discursive categories, to say nothing of the laws of literary genre.[2] This can be seen as much in the strange mixed genre of his own autobiographical history texts (*Mont-Saint-Michel and Chartres* and *The Education*) as in his ironic treatment of the presumably pure disciplines out of which he created such heterogeneous and heterodox categories as "Conservative Christian Anarchism."[3] I will have something more to say about Adams, but first it is necessary to look more closely at the relations between Pound and Whitman as a crisis of moment for modernism and Americanism in which current literary theory, itself riddled with generic and genealogical anxieties, still has a stake.

Early and late, American writers have sung of themselves as additions to or parasites upon an unbroken legacy of Western Art—marking both its completion and its unaccountable excess. Thus, not without quoting Whitman, Pound claims his American heritage and more: "I am (in common with every educated man) an heir of the ages and I demand my birthright. Yet if Whitman represented his time in language acceptable to one accustomed to my standard of intellectual-artistic living he would belie his time and nation. And yet I am but one of his 'ages and ages encrustations' or to be exact an encrustation of the next age" (SP, 145). Whitman is honored there, as Pound would always honor him, for faithfulness to his time and language; like Dante in Italy and Chaucer in England, Whitman wrote in his native or vulgar tongue and thereby created the possibility of a new poetry if not of an American language.

Pound is torn between refining the American, the Whitmanesque, idiom and acquiring greater erudition and a more impressive ancestry. At the same time that he complains of Whitman's inadequacies, Pound approves his nativist project. Yet he does so by translating it into a series of metaphors—the organic and the destructive, the archeological and the archeoclastic, the genetic and the accidental—for that American mythos of history that has always been similarly befuddled. Thus Whitman is the encrustation upon as well as the founder of a still uncertain "American tradition" to which Pound turns in order to overturn. And Pound is part of the new which is an encrustation upon both the last and the next ages, only a moment of transition between two disjunct poetic and genealogical lines.

However "congenial" or genealogically amenable Pound insists he is to the more recognized precursors of the Eliotic tradition, when he equates Dante and Whitman the earlier poet necessarily suffers a diminution. And, more importantly, the poetic hierarchy undergoes a leveling in which the privileged category of poetry hardly remains intact. Pound both acknowledges and represses these difficulties in charting an ancestry which is, if geometry and genealogy permitted such aberrations, a series of intersecting parallel lines of poet-hybrids who cut across several languages, artistic genres, and historical periods.

Just so, "textbooks" such as "How to Read" and *ABC of Reading* privilege those writers who worked radical innovations *within* the very tradition Pound seems to propagate. Even as early as "I Gather the Limbs of Osiris" (1911–12), he proposes to anthologize and canonize the "'donative' author [who] seems to draw down into the art something which was not in the art of his predecessors. If he also draw from the air about him, he draws latent forces, of things present but unnoticed, or things perhaps taken for granted but never examined" (SP, 25). Such an author/reader—let alone the modern reader who discovers and repositions him—automatically destabilizes the tradition by changing his inheritance and his legacy, that is, the "Art of his predecessors."

Whitman, the acknowledged inventor of "free-verse" and of the "democratic epic," necessarily disrupts poetry's old generic categories. And Pound, while always dissatisfied with Whitman's characteristic lack of refinement, tended to praise his forebear for such violations and to affect

such crudities in order to disturb the genteel tradition and inscribe his
personal signature of the prodigal modern. Further, he proposes separating Whitman's meaning from his verse, thus performing an operation
which rests on the assumption that *Leaves of Grass* is not an organic
poem; it does not, unlike his favorite Provençal poems, require full quotation or "direct presentation." According to Pound, this is because Whitman was so open to history and to the nonpoetic that *Leaves of Grass* is
as much history as it is poetry. If anything, Pound wants Whitman to be
less "poetic"—at least along the old lines—and more iconoclastic.

Indeed, it is Whitman the reader of culture and the scourge of tradition
Pound adopts as his American ancestor as, for instance, in *ABC of Reading,* when he says: "If you insist, however, in dissecting his language
you will probably find out that it is wrong NOT because he broke all of
what were considered in his day 'the rules' but because he is spasmodically conforming to this, that or the other . . . using a bit of literary
language" (ABCR, 192). However uneasy Pound might be about adopting the style of *Leaves of Grass* for his own poetic or pedagogical strategies, he never ceases acknowledging Whitman's *critical* (both crucial
and interpretative) function of turning the European poetic tradition toward the broader area of (American) culture. As we will see, the replacement of "tradition," a word that must be associated with Eliot's fabrication of his conservative modernism, by "culture," and finally Pound's
neologism, "Kulchur," graphically as well as semantically marks a disruption of the generic and genealogical orders.

Of his own relationship to Whitman, Pound says, "Personally I might
be very glad to conceal my relationship to my spiritual father and brag
about my more congenial ancestry—Dante, Shakespeare, Theocritus,
Villon, but the descent is a bit difficult to establish. And, to be frank,
Whitman is to my fatherland (*Patriam quam odi et amo* for no uncertain
reasons) what Dante is to Italy and I at my best can only be a strife for
a renaissance in America" (SP, 145–46). From a certain perspective,
Whitman occupies the position of first poet, the flower of that belated
and imported "American Renaissance" that originated in Emerson, or even
the position of the disseminator of an American, in contrast to the English, poetic language.[4] By these accounts, he is no more than an inter-
loper in Europe, especially the idealized Europe of pre-Renaissance Italy
where Pound would situate himself and his American ancestor. Whitman
is both the origin of American poetry and its unaccommodated original;
perhaps because he was thus, like Americans generally, obsessed with
origins yet compelled to be original in the exercise of individuality. At
least Pound seems to say as much when he is at pains to decide where

Whitman belongs in the greater scheme of things—as well as in the recent history of poetry.

Faced with a similar dilemma in the figure of that evolutionary glitch, the ganoid fish, *Pteraspis,* a still extant Paleolithic species passed over by evolution, Henry Adams observed that "to an American in search of a father, it mattered nothing whether the father breathed through lungs, or walked on fins, or on feet" (E, 229). Surely Pound's choice of precursors is not quite that indiscriminate. Whitman's centrality to a nativist poetics is hardly in doubt; nevertheless, he resists incorporation into an organic tradition or literary history as fiercely as Adams's fish resists Darwin's neat taxonomy.

Yet Whitman himself had deliberately reorganized more than one category within the poetic tradition and the larger province he named "American culture," a recurrent phrase, if not a new and privileged Idea, in his criticism as well as his poetry. In terms Pound would use in his poetry as well as his "culture criticism," Whitman had tried to redefine "culture" and his own European cultural inheritance. Not surprisingly, "culture" was a troublesome word for the egalitarian "rough" whose poetry and polemics were deployed against the narrow definitions of poetry and art advanced by "gentlemen of culture." In "Democratic Vistas," for example, Whitman asks how an American culture might be cultivated.[5] One should note that his analysis proceeds by a series of agricultural metaphors resting on a pun:

> The writers of a time hint the mottoes of its gods. The word of the modern, say these voices, is the word Culture.
>
> We find ourselves abruptly in close quarters with the enemy. *This word Culture, or what it has come to represent, involves, by contrast, our whole theme,* and has been, indeed, the spur, urging us to engagement.
>
> Certain questions arise. . . . *Shall a man lose himself in countless masses of adjustment, and be so shaped with reference* to this, that, and the other, *that the simply good and healthy and brave parts of him are reduced and clipp'd away, like the bordering of a box in a garden?* You can cultivate corn and roses and orchards—but who can cultivate the mountain peaks, the ocean, and the tumbling gorgeousness of the clouds? Lastly—is the readily given reply that *culture only seeks to help, systematize, and put in attitude, the elements of fertility and power,* a convulsive reply?
>
> *I do not so much object to the name, or word, but I should certainly insist, for the purposes of these States, on a radical change of category,* in the distribution of precedence. (DV, 479; emphasis added)

Like Whitman, Pound felt uneasy about the limited and privileged term "culture," which he figures as both a continuum and a wearing down of classical unity into indistinct fragments. Here is one of his most quoted definitions of "culture," one which is nonetheless not granted the complexity and inconsistency it clearly presents: "European civilization or, to use an abominated word '*culture*' can best be understood as a *medieval trunk with wash after wash of classicism going over it.* That is not the whole story, but to understand it, you must think of that *series of perceptions, as well as of anything that has existed or subsisted from antiquity*" (ABCR, 56; emphasis added).

Pound has again resorted to catachresis, to mixed metaphors that jumble the classical inheritance with subsequent revivals or "perceptions" of it. In a fiction that does not resolve as smoothly as Eliot's "Tradition and the Individual Talent," culture is here imaged as both the perennial family tree and as a shore or even a painting effaced by waves or washes of color. Like "encrustations of the next age," washes cover over and, at least to all appearances, change what is underneath. In the case of whitewash or of watercolor washes, such simple distinctions as object and covering can become completely obscured. Furthermore, in Pound's model, "classicism," which should by rights precede the "medieval," comes later, as a series of additions to, rather than a revelation of, previous cultures. Thus belated, American literature which began in New England or Brooklyn or, as Pound claims, in Virginia, was in a position to work a renaissance or a rejection of the old cultural forms—or perhaps to have it both ways, as Emerson and to some extent Whitman hoped.

In speaking of an American culture, in repeating once again that American gesture of mapping a national history, Pound suggests that America should open art and literature to the nonliterary and the inartistic. In "The Jefferson-Adams Letters as a Shrine and a Monument," for instance, he insists that American literature originated or culminated in the fugitive journals and correspondence of John Adams, Jefferson, Franklin, and Van Buren. Pound denies any fall away from classicism and thereby endorses the heterogeneity characteristic of America and modernism, thus: "Our national culture can be perhaps better defined from the Jefferson letters than from any other three sources and mainly to its benefice" (SP, 148–49).

Jefferson and the other founding fathers are praised for repeating the ordered multiplicity Pound had uncovered in Dante. With regard to the presidential correspondence, he recalls Dante's phrase, "'*in una parte piu e meno altrove*'" (SP, 150), which he uses to suggest that Jefferson and Adams were polymaths who nevertheless maintained a sense of proportion and a scale of values. In describing the "palimpsest" of their "Mediterranean state of mind" as the best "intellectual filing system," Pound admits that the world and works of his favorite Americans contained "things neither perfect nor utterly wrong, but arranged in a cosmos, an order, stratified, having relations one with another" (SP, 150). Despite the fact that here and in the "Jefferson/Adams Cantos" Pound tends to focus on those details that refuse incorporation into any sort of organic whole, and on the infinitely interpretive relations among writers and writings rather than on monolithic ideas, he memorializes the Jefferson/Adams letters in the following terms: "The MAIN implication is that they stand for a life not split into bits. Neither of these two men would have thought of literature as something having nothing to do with life, the nation, the organization of government" (SP, 152). Pound's category of American culture or the American "cosmos" (the unacknowledged adoption of the Whitmanesque term startles) is large enough to accommodate a great deal more than traditional art and literature, however minuscule Pound's quotations from the text of America eventually become.

Not without contradiction, then, Pound endorses the thrust of Whitman's change in the category "Culture," and thus tacitly approves the democratic and even the antipoetic elements of *Leaves of Grass*. Rather than smooth Whitman's roughneck image or remake his poem into a part of an epic continuity, Pound employs the earlier American as laborer against the conservative and elitist aesthetic which he nevertheless hoped to export intact back to America, as that "high modernism" he is credited with founding. His motives as well as his metaphors are mixed, and his translation of the European into an American culture is neither direct nor untroubled. Playing more on the architectonic than the organic, and in the name of Whitman as well as the names of more accepted figures, Pound issues a sort of Americanist manifesto: "It seems to me I should like to drive Whitman into the old world. I sledge, he drill—and to scourge America with all the old beauty . . . and with a thousand thongs from Homer to Yeats, from Theocritus to Marcel Schwob" (SP, 146).[6]

In his proposed destruction and reconstruction of the old beauty, Pound equates Whitman, Homer, Yeats, and the historian Marcel Schwob in a list he would supplement and reassemble throughout his literary career, adding, as time went on, writers from disparate ages and discourses. This was one strategy by which he would, as he said, "make all ages contemporaneous." But the cost of this ahistoricism or cultural relativism is a destabilizing of the poetic as well as the political hierarchies and privileges he never ceases claiming for poetry, the acme of culture by his own definitions.

Pound wanted both the stability of cultural monuments (though hardly Matthew Arnold's "touchstones") and the untamable action of new dis-

coveries and unusual methods. Frobenius's *Kulturmorphologie* and specifically the figure of "paideuma" satisfied Pound's conflicting desires
for novelty and respectability. Under various names, "paideuma" runs
throughout *Guide to Kulchur* and his other writings on American history.
At one point, he says: "The history of a culture is the history of ideas
going into action" (GK, 44). Later, in contrasting Frazer's mythography
to Frobenius's "archeology," Pound suggests that "paideuma" violates
chronology in his favorite way; it turns the distant past into a new prospect: "His [Frobenius's] archeology is not retrospective, it is immediate.
. . . To escape a word or set of words [culture and tradition] loaded up
with dead associations Frobenius uses the term Paideuma for the tangle
or complex of the inrooted ideas of any period" (GK, 57).

It is not going too far to claim that Pound enlisted the quirky German
anthropologist for his ongoing Americanist project of troping historical
and cultural "retrospection" into nativist "prospect," or even a prospec-
tus. After all, it was Emerson, that original American reviser, who began
Nature with the familiar admonition: "Our age is retrospective, it builds
the sepulchres of the fathers. It writes biographies, histories, and criticism. The foregoing generations beheld God and nature face to face; we
through their eyes. Why should not we also enjoy an original relation to
the universe" (*Complete Works,* Vol. No. I, 3). Pound articulated a similar concern about immediacy and originality in terms of original interpretations of the whole of culture. Yet for all his efforts at erecting original and somewhat paradoxically living and inrooted "monuments" to
Jefferson, Adams, Homer, Dante, *et al.,* Pound more than once echoes
Emerson in defining his own project: "The reader, who bothers to think,
may now notice that in the new paideuma I am not including the monumental, the retrospect, but only the pro-spect" (GK, 96).

In spite of himself, Pound proposes to undermine, if not to dismantle,
those artistic structures which had never been successfully transplanted
onto American soil. This is to say that when Pound supplements American poetry with European beauty, he exaggerates the stability and coherence of the American as well as that of the other order he calls tradition. Yet he neglects to note Whitman's ambivalence toward the European
tradition and thus he forgets the admixture of traditionalism and iconoclasm already stamped into "the American." His "Feelings About Walt
Whitman" are in fact characteristic of the schizophrenic loyalties the
American writer has always felt toward the Old and New worlds. A longer
essay might consider other segments of the broken and often subterranean
line that stretches from Emerson to those postmodern writers who still
answer "The American Scholar's" call to an American poetics. But in-

stead, I would like at this point to make a kind of European and post-modern detour, in order to call upon an even more recent commentary on what we have come to call American modernism. I will refer to a programmatic essay by Gilles Deleuze and Felix Guattari, entitled "Rhizome."[7]

"Rhizome" was undertaken to explain the Frenchmen's micro-political assaults upon such entrenched conventions of Western philosophical and literary discourse as the single author, the unified book, and the segregation of disciplines within the academy. The essay opens, for instance, with Deleuze and Guattari positioning—or, better, "deterritorializing"—themselves in the midst of an ongoing cultural exchange that will not admit of *their* own uniqueness. Their very insistence on shared and unaccountable authorship is an embarrassment to literature, psychology, and philosophy, all disciplines that consider individual consciousness primary to systematic and serious discourse. They began thus, conscious of a belatedness that includes their own past writings:

> We wrote *Anti-Oedipus* together. As each of us was several, that already made quite a few people. . . .
>
> A book has neither subject nor object; it is made up of variously formed materials, of very different dates and speeds. As soon as a book is attributed to a subject, this working of materials and the exteriority of their relations is disregarded. A benificent God is invented for geological movements. In a book, as in everything else, there are lines of articulation or segmentation, strata, territorialities; but also lines of flight, movements of deterritorialization and of destratification. (OL, 1–2)

A similar and even more anxiety-ridden sense of belatedness has characterized American writing from Emerson to the present, and, as we have seen, applies even to Whitman, despite the critics' tendency to accept his immediate and "barbaric yawp" as a return to the primal voice. Marking this self-consciousness, which wants to preserve a genealogical sense of order yet claim a sense of individual identity in its own disruptiveness, is a gesture that Pound whimsically called, in a letter to William Carlos Williams, "the American habit of quotation" (L, 124).[8] Quoting and stealing, the American writer from Emerson to Poe on would "deterritorialize" his past. But this also meant subverting any claim to originality and authority over his own text.

Indeed, Pound calls into question his own authority if not his own authorship, as he persists in valuing the unstoppable exchanges of subjects and objects within the culture and its adequate history:

> We do NOT know the past in chronological sequence. It may be convenient to lay it out anesthetized on the table with dates pasted on here and there,

but *what we know we know by ripples and spirals eddying out from us* and from our own time.

There is no ownership in most of my statements and *I cannot interrupt every sentence or paragraph to attribute authorship* to each pair of words, especially *as there is seldom an a priori claim even to the phrase or half phrase*. (GK, 60; emphasis added)

I do not wish to exaggerate the similarities between Pound and the two French writers who have no reason to think like the traditionalist/modernist/Fascist writer, whom they instead have every reason to reject. Nevertheless, Pound, and Whitman as well, pushed against the limitations of books and logical discourse in much the same way as Deleuze and Guattari propose to do:

A tiresome feature of the Western mind is that it relates actions and expressions to external or transcendent ends, instead of appreciating them on the plane of immanence according to their intrinsic value. For example, insofar as a book is made of chapters, it has its culminating and terminal points. What happens, on the contrary, with a book made of plateaus, each communicating with the others through tiny fissures, as in the brain? We shall call a "plateau" every multiplicity connectable with others by shallow underground stems, in such a way as to form and extend a rhizome. We are writing a book as a rhizome. (OL, 50)

If such radical deployments of the problems inherent in writing remain—or have again become—surprising to the American critical ear, questions such as "What is an author?" "What are the ends of philosophy?" and the like have become thematic commonplaces in recent Continental thought. Yet one might claim that they are native to American poetics and certainly to Pound's version of the modern, both of which unsettle categories of the Western literary tradition. Although they make no reference to Pound's modernist interventions, Deleuze and Guattari suggest as much when they say that Western culture, which they trace back to its roots in agriculture, has always privileged trees, if in a most ambivalent fashion: from the Edenic tree of knowledge to the family tree of Indo-European languages referred to in Chapter Five. Borrowing a distinction from botany and amplifying its metaphoric resonance, they oppose the arborescent or treelike culture to their own rhizomatic or weedlike writing. And, most important here, they find their precedent in America's weedlike multiplication of influences and tangled roots.

To summarize briefly the botanical definitions: trees have fixed roots and structural permanence; rhizomatic plants, which include most tubers, orchids, and such virulent weeds as the Virginia creeper, lack fixed roots and instead have filamental shoots that grow horizontally just beneath the surface of the ground and sometimes project aerial roots in all directions. Trees grow vertically; rhizomatic plants grow horizontally. With work, trees can be cut down, but rhizomes left in the ground can hide, hibernating and spreading subterraneously until they eventually give rise to new plants or more weeds, often in surprising locations and configurations. By extension, and to quote Guattari and Deleuze, "Arborescent systems are hierarchical systems comprised of centers of significance and subjectivization, of autonomous centers. . . . It is curious how the tree has dominated Western reality, and all of Western thought, from botany to biology and anatomy, and also gnosticism, theology, ontology, all of philosophy . . . : the root-foundation, *grund, fondements*" (OL, 36, 40).

Rhizomes can, however, masquerade as trees. Sometimes, as is the case with Pound and Adams, the grafting of a family tree of poetry and the other disciplines becomes a subtle parody of the search for origins. Or, as in Whitman, one cannot find the trees for all the individual, heterogeneous, and heteroclite leaves of grass and of paper. Curiously enough, Deleuze and Guattari claim that "American literature," notwithstanding its perennial search for origins and originality, is the exemplary rhizomatic text, both for its origins in European and its writings about American culture. If the Frenchmen might be said to give "America's rhizomatic literature" a univocity we all know it lacks, they nevertheless expose the tangled themes of the Americanist problematic. I quote one particularly stunning segment:

> America should be considered a place apart. Obviously it is not exempt from domination by trees and the search for roots. This is evident even in its literature, in the quest for a national identity, and even for a European ancestry or genealogy. . . . Hence the difference between an American book and a European book, even when an American sets off pursuing trees. A difference in the very conception of the book: "Leaves of Grass." Nor are directions the same in America: the East is where the arborescent search and the return to the old world takes place; but the West is rhizomatic, with its Indians without ancestry, its always receding borders, its fluid and shifting frontiers. (OL, 42)[9]

Notwithstanding the rather complex and foreign, not to say subversive, character of Deleuze and Guattari's argument, it would seem that much of American writing has been involved with rhizomes, that is, with the wild weedlike growths that undermine the organic or agricultural order so fundamental to Western thought and its literature. Emerson, the putative founder of what is now recognized as a de-centered—if not a rhizomatic—American poetic tradition, has a good deal to say about the American underbrush which refuses to sustain trees or other rooted structures. One of his most striking and heuristic uses of the botanical metaphor takes the form of a self-conscious complaint against the very diffuseness and errancy epitomized by his own essays, journals, and miscellaneous notes. In a journal passage dated 1847, he says, "Alas for America as I must so often say, the ungirt, the profuse, the procumbent, one wide juniper, out of which no cedar, no oak will rear up a mast to the clouds! It all runs to leaves, to suckers, to tendrils, to miscellany."[10]

Guattari and Deleuze were probably not thinking of Emerson at all, nor were they embarrassed by Whitman's *Leaves of Grass* in quite the same way as were modern American poets like Pound, Eliot, and Williams. Yet what in "Rhizome" appears as a privileged, cross-discursive miscellany and the projection of a subterranean American countertradition (poetic counterculture?) is related to the ambiguous nationalism of Pound's modernist program. Moreover, as we have seen with regard to Whitman's *sledging* and his *sap* of nativism, Pound's thoughts about American poetics are frequently conveyed in strained botanic metaphors.

Pound certainly set out pursuing trees, if with the quixotic or at least the unnatural aim of grafting an ancient European genealogy onto the new growth of American literature to which Whitman was both precursor and product. Despite continual efforts to repair the rootlessness of *Leaves of Grass,* Pound's own long poem, and his criticism both in and out of that poem, share much with the text Deleuze and Guattari somewhat cryptically identify as representative of "the American rhizome." Like Whitman's poem, *The Cantos* is heterogeneous, insistently incomplete or provisional; composed of fragments of various historical themes and poetic forms, it too changed over the history of its writing. Indeed, the further back in time and the farther east Pound ventured in search of his "legitimate ancestry," the more fragmentary his poem became; that is, the more history Pound included in his poem, and the more ancestors and antecedents he collated into metaphors and models of his tradition, the more insistent became his questioning of the fundamental concepts and figures comprising any nostalgic quest for origins. Perhaps this contradiction is most pronounced in "The Chinese Cantos," which records four thousand years of dynastic history in genealogical tables which Pound interrupts with distracting and often idiosyncratic programmatic statements. This textual layering is conveyed by a mixture of organic, economic, and architectural metaphors, comprising at best simulacra of System or tradition.

"The Chinese Cantos," published with the "Jefferson/Adams Cantos" and obsessed with order and systematization, is composed of virtually unreadable fragments translated through French transliterations of often corrupt and incomplete records written by thousands of court historians whose interested selections can hardly be considered the work of objective or accurate historiography; nor can their various styles and themes be gathered into any sort of coherent book, let alone an organic American epic. Nevertheless, Pound was expressly attempting to use these texts to establish a general economy or cultural paradigm that could in some way mark a continuity between the ancient agrarian civilization of China and such later idealized cultural repetitions as Jeffersonian democracy. Despite the efforts of sympathetic critics to translate Pound's stated intention into an organic poem or an epic held together by such devices as those dubbed "subject rhymes" and "nodal repetitions," both in theme and form the text remains a kind of native hybrid (to use an oxymoron), exhibiting the sort of rootlessness and excess Deleuze and Guattari call rhizomatic.[11] Perhaps more than any of the Cantos except for *Rock Drill,* the Chinese sequence exceeds even the "wash upon wash of classicism" that Pound both charted and descried. Moving away in both directions from European Classical culture (east and backward in time to China as well as west and forward in time to America), *The Cantos* is rhizomatic, that is, not rootless but, as Pound said, "the tangled and inrooted ideas of any age."[12]

Two examples from "The Chinese Cantos" will suffice to show that Pound's own self-conscious metaphors undercut the organic or, more precisely, the arborescent tradition he set out to recover and, as it turned out, to cut and graft into a different sort of growth. Canto LII reintroduces and augments the series of ostensibly architectonic metaphors and mythological systems that weave through *The Cantos* like tendrils, surfacing only as new sources and themes are introduced or old ones modified. At this point, Pound signals a turn from sixteenth-century Venetian economics back to the mythic founding of agriculture in prehistoric China. Preceding the Canto proper, one of the poem's few prose notes announces the theme of the two decads which encompass Chinese history and the Jefferson/Adams period of America. Here, Pound feels compelled to reflect on what impedes or disrupts his own elucidation of this linkage.

After a table of contents more appropriate to a textbook like *Guide to Kulchur* than to an epic, he makes his excuses in somewhat misleading advice to the reader:

Note the final lines in Greek, Canto 71, are from Hymn of Cleanthes, part of Adams' *paideuma:* Glorious, deathless of many names, Zeus eye ruling all things, founder of the inborn qualities of nature, by laws piloting all things. (LII: 256)

On the one hand, he insists that the Eleusinian or agricultural Zeus serves as a principle of order and exact repetition; on the other hand, he underscores the problems inherent in translating proper names and ideas from one culture into another and, by extension, the difficulties of substituting one set of organizing or textual metaphors for another. Rather than mutually translatable or originary ideas of order, Pound offers a series of mixed metaphors, which gain complexity as he traces them back through time and across cultures: nature vies with Classical poetry, agricultural myths with avant-garde art, for preeminence in an "American paideuma," which names the roots and shoots that form textual tangles rather than the permanent roots and organized structures of (family) trees. While it seems to retain a grounding in nature or vegetation myths, Pound's cultural history uses a different organic/botanic model. His repeated efforts at "clearing the underbrush" and breaking ground aside, Pound seems always to focus on the wayward and weedlike proliferation of words and of translations.

In *The Cantos'* American *paideuma,* historical particularity and clear definitions give way to heterogeneity and lawlessness that might permit Pound to graft independent and incompatible systems, to the point where it is no longer possible to distinguish a main stem from dependent offshoots. In the cutural exchange between revolutionary America and Imperial China, John Adams, like Pound's rather idiosyncratic Zeus, figures centrally in the founding of both governmental and agricultural systems. Adams and Jefferson are praised for erecting monuments, including statues of Washington, and for valuing the active and revolutionary over the static and monumental. By way of reconstructing viable poetic and economic programs for modern America—or creating them for the first time out of fragments of an old "American Dream"—Pound advances a wayward textual economy of multiplying references or, in his phrases, "spermatic thought" and "interpretative metaphors." His text embraces as many contradictions, and includes as many senses of "culture," as Whitman's.

As though to compound the disorder of his various systems, and to deny the fundamental principle of Imagism by which he rejected rhetorical ornament, Pound claims that his use of Chinese ideograms and all the other foreign quotations is almost purely supplementary and ornamental. In "The Chinese Cantos" one suspects that Pound's simply erroneous translations and transliterations border on irony, since he seems to know more about the complexities of cultural exchanges and translations than he practices. In the note preceding Canto LII, he seems to warn and yet to reassure the reader, aware at least that he is complicating linguistic and cultural matters: "Foreign words and ideograms both in these two decads and in earlier cantos enforce the text but seldom if ever add anything not stated in the English, though not always in lines immediately contiguous to these underlinings" (LII: 256).

Indeed, for most readers, the Chinese glyphs and Pound's other interpolations add little to the English, though they are likely to erupt weedlike into the English text at places where their meaning remains indeterminate. Not only do Pound's foreign usages violate Imagist doctrine, but he also seems willing to court uncontainable metaphoricity, not to say rhetorical ornament, for the sake of inclusiveness and, perhaps, in order to appear cultured in a random sort of way. Chinese characters often stand as signs of the exotic, as metonyms for incompatible or irreducible systems, and, more to Pound's purpose, as impressive monuments to genealogical and State order. In this way, Pound complicates recurrent Classical references by interpolating into his text less congenial precursors of his ideal America—even in the very name of the Adams family line, and using Brooks Adams's notion of cultural/economic exchange.[13] Pound is often self-conscious and always deliberate about his unorthodox procedures for fabricating what he will claim is an organic tradition.

The very claim that his text can be easily domesticated, that foreign borrowings contribute emphasis rather than meaning, involves Pound in a play of figures, a questioning of language, that he could not contain. Rather than underlining significant connections or tracing subterranean root systems among the various languages, arts, and sciences he wanted to appropriate, his Chinese and other foreign inscriptions radiate tangled lines of force and influence in all directions, jumping from plateau to plateau and following the "lawless laws" of rhizomes.

One of the more suggestive instances of these fertile outcroppings occurs in Canto LIII, where Pound's nostalgic record of the agricultural and monetary reforms instituted by the founder of the ancient Shang dynasty is punctuated by that frequently cited modernist imperative, "Make it New." I quote: "In 1760 Tching Tang opened the copper mine/ (ante Christum)/ made discs with square holes in their middles/and gave these to the people/wherewith they might buy grain. . . . Tching prayed on the mountain and/wrote MAKE IT NEW/on his bath tub./Day by day make it new/cut the underbrush, pile the logs/keep it growing" (LIII: 264–65). Pound's free translation of the imperial motto, which had been coined by Confucius roughly a millennium after Tching's reign, performs the same sort of destructive and/or reconstructive (one could nearly say de-constructive) task for which the latter-day American enlisted Whitman as drill and sledge. Thus, even in this most radical effort to affirm genealogical order, an absolute origin and continuity to culture, Pound uncovers the ubiquitous conflict of tradition and innovation in which history, science, and poetry must begin over again—and differently—every day.

Pound was compelled to reflect on the plethora of masterpieces, discourses, and traditions which lend to his poem a chaos of systems that cannot be reduced to a master system. Often his self-reflections on this palimpsestic structure involve an uneasy mixture of the organic and the architectonic; his reflections become inextricable from the textual machine he had set in motion. Especially in the later Cantos, Pound addresses what we have been calling the American rhizome as against European and Asian arborescence. For instance, in Canto LXXXV, the first poem of *Rock Drill*, where he recalls the natural basis of sound economy and right government from the Shang dynasty to the present, Pound refers to the sacred ash of Norse mythology: "From T'ang's time until now/That you lean 'gainst the tree of heaven/and know Ygdrasil" (LXXXV: 545).

A few pages later in the same Canto, however, Pound tries to translate the myth of an originary unity, or transplant a rooted tradition, into the hostile climate of America. By his account, his native culture has "No classics,/no American history/no centre, no general root" (LXXXV: 549). This particular turn or translation back into the American literary scene ends one of *The Cantos'* many catalogues of those nonliterary disciplines (including philology, geology, mathematics, biology, and publishing) from which Pound, rather like Henry Adams, borrowed tags and the proper names of culture heroes, though not a system. Moreover, his very complaint places him in a certain pattern of anxiety reaction to America's decenteredness and miscellaneousness. As we have seen, Emerson, whose writings epitomize the very tendencies they dismiss, was already in this tenuous and, it would seem, American position.

Finally, in "Draft of Canto CXIII," Pound's own system building culminates in an unsteady equation of three metonyms (rather improperly used proper names of the key figures from disparate arts) associated with the unrelated though internally coherent disciplines of music, taxonomy, and genetics. He depicts Paradise thus: "Yet to walk with Mozart, Agassiz and Linnaeus/'neath overhanging air under sun-beat/Here take thy mind's space/And to this garden" (CXIII: 786). In this way, Pound's poem—for which a baroque gallery (a bust of Mozart here, a specimen chest in the style of Goethe's study there, and, in the midst of all these monuments, the artist creating new masterworks at an easel or writing desk surrounded by discarded drafts) would be a more fitting metaphor than even that wild mental landscape or garden of the muses—collects fragments, tags, and whole strata of traditions and discourses.

The poet was not unconcerned about his tendency to miscellany. After all—and *after all*—he says, in the fragment of Canto CXVI, "I am not a demigod,/I cannot make it cohere. . . . Disney against the metaphysicals. . . . a nice quiet paradise/over the shambles" (CXVI: 796). If we take seriously Pound's sometimes tragic and sometimes more strategically ironic comments about the incoherence of his poem and of the cuture which willy-nilly gave rise to it, we come to recognize that he poses fundamental challenges not only to poetry but to virtually all those Western scientific and metaphysical traditions that have become the targets of current poststructuralist readings in the texts of Deleuze, Guattari, and certain wayward tendrils of recent American criticism.

Yet this deconstructive strain, this self-conscious tendency to subvert self-reflection, has always been present at least on the margins of American literature. No writer has been more crucial in this regard than Henry Adams, though one hesitates to name so insistently marginal a figure as central to America's critical—let alone its literary—tradition. As he visited the cultural shrines of Europe, Adams insisted that he was a tourist and an uncle, neither an expert nor the father of a new historiography or a new critical method. He discovered neither a continuous line of descent nor a completed circle of cultural achievements; instead, he multiplied genealogical anxieties until his revision of European culture came to parody itself in a swirl of interpretations launched against the persistent desire for closure—at least on the model of the hermeneutical circle or the autotelic text. His choice of literary and architectural monuments at least anticipates Pound's, even as it follows Hawthorne's *The Marble Faun* and a whole line if not a literary genre of travel romances that trace the prehistory of American culture.[14] Adams was perversely fond of digging up problems such nostalgic texts are generally committed to avoiding. Meditating, for instance, on the edifice of Mont-Saint-Michel, Adams recalls several conflicting accounts (some in tapestries, some in poems) of the Battle of Hastings from which England dates its nationhood and about which the record Mont-Saint-Michel tells another version. Not only does he underscore the tangled heritage of England, and thus his Anglo-American (or is it Franco-American?) inheritance, but he also shows that the different interpretations of historical, architectural, and literary texts can make cultural and artistic hierarchies tremble. He is at once annoyed and motivated by an anarchical scepticism, one that in a particularly American fashion refuses to be destructive or simply affirmative as it finds metaphors and interested interpretations where fundamental truths were supposed to have existed:

> But the feeling of scepticism, before so serious a monument as Mont-Saint-Michel, is annoying. The "Chanson de Roland" ought not to be trifled

with, at least by tourists in search of art. One is shocked at the possibility
of being deceived about the starting point of American genealogy. Taillefer
and the song rest on the same evidence that Duke William and Harold and
the battle itself rest upon, and to doubt the "Chanson" is to call the very
roll of Battle Abbey in question. The whole fabric of society totters; the
British peerage turns pale. (MSMC, 20)

Yet doubt he would, and annoy his fellow Harvard historians with sceptical analyses of facts and the empirical historicism by which culture and
the curriculum established their heritage and legitimacy. Rather than defer
to "science"—or to the interpretative metaphors Pound claims to have
borrowed from science—Adams brings his criticism to bear on the modern sciences and philosophy as simulated unities or systems to which the
arts have mistakenly turned. Adams uncovers the progressive complexity—indeed the original heterogeneity—that preceded the building of the
first stratum of such venerable textual and architectural palimpsests as
Chartres and Mont-Saint-Michel.
  Since the full implications of the critical and parodic thrust of Adams's
writing has been glossed over by those readers who appreciate his genteel
prose and cultured sensibility more than his destructions of ideal geneal-
ogies, it is worth noting his manner of reconstituting a texture or tissue
of fragments. He does not lament the ruined state of the churches and
other "sacred places" of Medieval history. Quite the contrary, he takes
advantage of the opportunity: using the flexible medium of his prose, he
takes a striking piece from the façade of, say, a Romanesque church and
transposes it onto an incomplete, not to say stylistically incompatible,
Gothic facade. By this method, which is anathema to history and impossible for architecture, his various fragments become figures in an ever-
growing construct that is grounded in complex synecdoches, in which
details of various styles represent whole ages that are stitched into Adams's unique prose patchwork that never resolves into a whole book or
even an exemplary beginning for a serious art historical study. The ideal
edifice he erects is more modern than Medieval, more a phantasmagoric
exchange of images than a nostalgic recuperation of primitive or organic
art:

  Here at Mont-Saint-Michel we have only a mutilated trunk of an an elev-
  enth-century church to judge by. We have not even a facade, and we shall
  have to stop at some Norman village—at Thaon or Ouistreham—to find
  a west front which might suit the Abbey here, but wherever we find it we
  shall find something more serious, more military, and more practical. . . .
  So, too, the central tower or lantern—the most striking feature of Norman

churches—has fallen here at Mont-Saint-Michel, and we shall have to re-place it from Cérisy-la-Forêt, and Lessay, and Falaise. (MSMC, 9)

Whether or not the modern poet had the proto-modern tourist in mind, the "mutilated trunk," to say nothing of its repair by further fragmentation and the mixing of genres into a sort of collage, anticipates Pound's poetic method. It is worth noting that, while medievalism and Provençal culture in particular were all the rage in Pound's college years, he was especially fond of "The Song of Roland" and granted to its writer and to such trou-badours as Bertran de Born a privileged place in his canon of innovators and other heretics. From *Personae* to *The Cantos,* he employed a method of construction that he claimed to have borrowed from Bertran de Born. We have already noted Pound's habit of carving out family trees and preferring washes of interpretation over rooted traditions; in this project of fragmentation, he goes a step further than Henry Adams (whom he fashioned the last of a long line of American politicians and active men whose lives were "whole"). He finds that the very poetics of the trou-badours involved the piecing together of fragments into anything but an organic whole or an unmediated song. In early long poems, "Na Audiart" and "Near Perigord," for instance, Pound constructs a "composite poem" on the model of the "composite lady of the troubadours," sent as tributes to "real ladies" and anthologizing the ideal features of renowned beauties immortalized by other poets. Thus, in American poems consisting of sometimes partially translated borrowings and stylistic parodies, Pound discloses that in poems written and performed at the same time that the great cathedrals and abbies were slowly being erected, there was no sim-ple relationship between art and life. No one planner, architect, or ob-server saw the completion of a cathedral, or even the final version of a poem in his lifetime. Furthermore, the most perfect songs were no more simple reflections of living women than Pound's new American poem, which was to be stitched together out of fragments already doubly textual because they were rooted in poems and not in Nature or Idea.

Pound's closest approach to Adams's architectural reconstructions is his admiration of Sigismundo Malatesta's *Tempio* (Cantos VIII–XI, *pas-sim*), that textual layering of pagan/Christian/Classical fragments which Pound called "both an apex and in a verbal sense a monumental failure" (GK, 159). While Pound could accept the heterogeneity of Malatesta's late-Medieval (re)construction, he considered similar American attempts, including William Randolph Hearst's San Simeon, dangerous and in-authentic. When it came to an American Renaissance, he could brook no variety, no individuality. For his version of American culture, Pound chose the rigid, if no less ersatz, center of Fascism. See, for example, his discovery of an Italian version of Jeffersonian agrarianism in *Jefferson and/or Mussolini*. The tangled roots of Pound's Fascism can be read in—or back into—American history. But this is a subterranean line we cannot follow out here.

If Pound was more troubled by his American heritage, his mixed inheritance, Adams was bemused by the prospect of the *New* overtaking the Old World. He too brought America with him, carrying his Adams and Quincy grandfathers from New England to his earlier continental ancestors—making, that is, a characteristically American version or reversal of the founding of Rome. In another passage that might be thought to adumbrate Pound's figure of the washes or waves of one culture over another, of modernism obscuring the clear demarcations of the European tradition, Adams confounds genealogy as well as history and geography. He discovers his native New England, as the trace of his own youth, even as he looks down under the foundation of Mont-Saint-Michel:

> From the edge of the platform, the eye plunges down, two hundred and thirty five feet, to the wide sands or the wider ocean, as the tides recede or advance, under an infinite sky, over a restless sea, which even we tourists can understand and feel without books or guides. . . .
>
> One needs only be old enough in order to be as young as one will. From the top of this Abbey Church one looks across the bay to Avranches, and toward Coutances and the Cotentin—the *Constantius pagus*—whose shore, facing us, recalls the coast of New England. (MSMC, 2)

In both *Mont-Saint-Michel* and *The Education* Adams catalogues historical adumbrations and repetitions in the self-reflexive manner usually associated with the most disruptive and defamiliarizing sorts of modernism. The organic intersects with the architectural, as the critic ironically restages the fall into complexity that made possible his project of recasting culture into something more exciting because less homogeneous and more clearly marked by interpretation. Adams opens art to science, but to a science that has always been fragmented and in a state of flux. According to Adams, science became baroque or rhetorical after Aquinas, and the baroque is still becoming. And it is American.

Centuries before Adams wrote his idiosyncratic architectural digest—that is, his autobiography of the growth of his own scepticism—there was no ultimate reference point for culture or the arts. Quite apart from his attraction to ruins and monuments, and the apparent nostalgia for the centering figure of the Virgin, Adams celebrates this multiplicity and dynamism. As he digs through layers of supplements to the ancient walls, he remarks the beginnings of the suspicion that organic unity was a mere dream. He suggests that modern science is only the natural—and we would have to say "rhizomatic"—outgrowth of a complexity as fundamental as the dream of organic art, which he criticizes as the "dogma" of orthodoxy and the central Idea. Of the birth of scientific scepticism out of Medieval theology, he says:

> Modern science, like modern art, tends in practice to drop the dogma of organic unity. Some of the medieval habit of mind survives, but even that is said to be yielding before the daily evidence of increasing and extending complexity. The fault, then, was not in man, if he no longer looked at science or art as an organic whole or as the expression of unity. Unity turned itself into complexity, multiplicity, variety and even contradiction. All experience, human and divine, assured man in the thirteenth century that the lines of the universe converged. How was he to know that these lines ran in every conceivable and inconceivable direction, and that at least half of them seemed to diverge from any imaginable centre of unity! (MSMC, 375)

American writers from Emerson to Whitman to Adams to Pound and beyond have naturally—or unnaturally—recognized this complexity and have nonetheless given life to a decentered tradition, if not a culture, called America or American literature or, more simply, modernism. Neither centering nor de-centering has been without its poetic and political dangers.[15]

NOTES

Chapter One

1. Here Pound invites—or dares—his readers to new (methods of) interpretations through Dante's words. Here and throughout I refer to the standard edition of Ezra Pound, *The Cantos* (New York: New Directions, 1972). I follow the convention of citing specific passages from *The Cantos* by Canto number in roman numerals and page numbers in arabic numbers—that is, XCIII:631 is "Canto XCIII," page 631. In his earliest criticism, he glosses these lines, *Paradisio* II: 1–2 (*O voi siete in piccioletta barca*), as an address in which the reader "is both warned and allured," *Spirit of Romance* (New York: New Directions, 1952), p. 142.

Subsequent quotations from Pound's published writings will be noted in the text as follows:

AA   *Active Anthology* (London: Faber & Faber, 1933).

ABCR   *ABC of Reading* (New York: New Directions, 1960).

CEP   Michael John King, ed. *Collected Early Poems of Ezra Pound* (New York: New Directions, 1972).

CWC   Ezra Pound, ed. *The Chinese Written Character as a Medium for Poetry*, by Ernest Fenollosa (San Francisco: City Lights Books, [1963]).

GK   *Guide to Kulchur* (New York: New Directions, 1970).

IM   Noel Stock, ed. *Impact* (Chicago: Henry Regnery, 1960).

J/M   *Jefferson and/or Mussolini* (New York: Liveright, 1970).

L   D. D. Paige, ed. *Selected Letters of Ezra Pound, 1907–1941* (New York: New Directions, 1971).

LE   T. S. Eliot, ed. *The Literary Essays of Ezra Pound* (New York: New Directions, 1971).

NPL   Ezra Pound, trans. *Remy de Gourmont's Natural Philosophy of Love* (New York: Boni & Liveright, 1922).

Per   *Personae: The Collected Shorter Poems of Ezra Pound* (New York: New Directions, 1971).

PVA   Harriet Zinnes, ed. *Ezra Pound and the Visual Arts* (New York: New Directions, 1981).

SP   William Cookson, ed. *Ezra Pound: Selected Prose, 1909–1965* (New York: New Directions, 1973).

SR   *Spirit of Romance* (New York: New Directions, 1952).

2. Friedrich Nietzsche, *Ecce Homo*, trans. and ed. with commentary by Walter Kaufmann (New York: Vintage Books, 1967), p. 264.

Subsequent quotations from the translated writings of Nietzsche will be noted in the text as follows:

BT   "The Birth of Tragedy," *The Birth of Tragedy and the Case of Wagner,* trans. Walter Kaufmann (New York: Vintage Books, 1967).

CW   "The Case of Wagner," *The Birth of Tragedy and the Case of Wagner,* trans. Walter Kaufmann (New York: Vintage Books, 1967).

EH   *Ecce Homo,* trans. and ed. with commentary by Walter Kaufmann (New York: Vintage Books, 1967).

GS   *The Gay Science,* trans. Walter Kaufmann (New York: Vintage Books, 1974).

NCW   "Nietzsche Contra Wagner," *The Portable Nietzsche,* trans. Walter Kaufmann (New York: Viking Press, 1954).

RW   "Richard Wagner at Bayreuth," *Thoughts Out of Season, Part I,* Vol. IV of *The Complete Works of Friedrich Nietzsche,* trans. Anthony M. Ludovici, ed. Dr. Oscar Levy, 2d ed. (New York: Macmillan, 1924). The first edition of this collection was published by F. N. Foulis (Edinburgh and London, 1909–1913). At the time Pound was contributing articles to *The New Age,* Ludovici was philosophy reviewer and Levy an occasional commentator on Nietzschean currents for the same weekly.

TF   "Truth and Falsity in the Ultra-Moral Sense," trans. Oscar Levy, *The Writings of Friedrich Nietzsche,* ed. Geoffrey Clive (New York: Mentor, 1965), pp. 503–15.

TI   *Twilight of the Idols and the Anti-Christ,* trans. R. J. Hollingdale (London: Penguin, 1968).

WP   *The Will to Power,* trans. Walter Kaufmann and R. J. Hollingdale, ed. Walter Kaufmann (New York: Vintage Books, 1968).

3. Martin Heidegger, *Nietzsche, Volume I: The Will to Power as Art,* trans. with notes and analysis by David Farrell Krell (San Francisco: Harper & Row, 1979).

Subsequent quotations from English translations of Heidegger's works will be noted in the text as follows:

BT   *Being and Time,* trans. John Macquarrie and Edward Robinson (New York: Harper & Row, 1962).

EGT   *Early Greek Thinking,* trans. David Farrell Krell and Frank A. Cappuzzi (New York: Harper & Row, 1975).

EP   *The End of Philosophy,* trans. with intro. by Joan Stambaugh (New York: Harper & Row, 1973).

IM   *An Introduction to Metaphysics,* trans. Albert Hofstadter (New York: Doubleday-Anchor, 1961).

N   *Nietzsche, Volume I: The Will to Power as Art,* trans. with notes and analysis by David Farrell Krell (San Francisco: Harper & Row, 1979).

PLT   *Poetry, Language, Thought,* trans. Albert Hofstadter (New York: Harper & Row, 1971).

QT   *The Question Concerning Technology and Other Essays,* trans. with an
intro. by William Lovitt (New York: Harper & Row, 1977).

4. This "New Nietzsche" is both the affirmative and the deconstructive Nietzsche
of contemporary rhetorical—and mostly French—critics, like Jacques Derrida,
Gilles Deluze and Philipe Lacoue-Labarthe, Paul de Man, Sarah Kofman, and
Michel Foucault. See David B. Allison, ed., *The New Nietzsche: Contemporary
Styles of Interpretation* (New York: Dell, Delta Books, 1977).

5. I mean to suggest the distinction between "text," (that is, writing as a play
of forces, including the writer's use of tradition, "pretexts," his or her place in
a critical dialogue and cultural milieu, his or her "subtexts," and the relation of
one's writings to others and to the world, or "contexts") and "books"—or, more
properly, "the idea of the book"—which is on the model of a self-contained
whole defined by generic rules, publishing, and other conventions. Such distinctions suggested as easily by Pound's fragmentary notes as more recent and
more rigorous examinations recall Derrida's equation of the questioning of the
book's privilege: "The idea of the book is the idea of a totality . . . which always
refers to a natural totality, profoundly alien to the sense of writing. It is the
encyclopedic protection of theology and of logocentrism against the disruption
of writing, against its aphoristic energy, and . . . against difference in general.
If I distinguish the text from the book, I shall say that the destruction of the
book, as it is now under way in all domains, denudes the surface of the text.
That necessary violence responds to a violence that was no less necessary." Jacques
Derrida, *Of Grammatology,* trans. Gayatri Chakravorty Spivak (Baltimore: Johns
Hopkins University Press, 1976), p. 18.

6. Pound clearly and frequently depreciates criticism, and all philosophical
discourse, in favor of poetry. Nevertheless, he sometimes claims to have felt that
his own prose could or should break with grammar, logic, and philosophy to
approximate the greater freedom of poetry. Here is a startling example of his
invocation of a certain poetic license; arguing for the "unavoidable *difference* in
truth itself" (J/M, 42; emphasis added), he opposes poetic tropology, rhetorical
tricks against logical discourse: "I am not putting these sentences in *monolinear
syllogistic arrangement,* and I have no intention of using *that old form of trickery
to fool the reader, any reader, into thinking I have proved anything*" (J/M, 28;
emphasis added).

7. While Heidegger frequently lends Nietzsche philosophical legitimacy by
etymological and other philological arguments that tend to defuse the revolutionary potential of his precursor's words, he also approximates Nietzche's word
fetishism in his own notion of bracketing. He both credits Nietzsche's disruptions
and seeks to neutralize them by *re-framing* them in proper (ahistorical, nonrhe-
torical) philosophical language: "[Nietzsche] knows, only as a creator can, that
what from the outside looks like a summary presentation is actually the config-
uration of the real issue, where things collide against one another in such a way
that they expose their proper essence. Nevertheless, Nietzsche remains under
way, and the immediate casting of what he wants to say always forces itself upon

him. In such a position he speaks directly the language of his times and of the contemporary 'science'. When he does so he does not shy away from conscious exaggeration and one-sided formulations of his thoughts, believing that in this way he can most clearly set in relief what in his vision and in his inquiry is different from the run-of-the-mill" (N, 50).

8. This "Amerikun" neologism—and neo-spelling as well as a "free translation" of (Frobenius's) *Kultur(morphologie)*—parodies elitist and British "culture" in a fashion initiated by Emerson and executed by Whitman. Its implications will be examined in detail in Chapter Five, its prehistory in the Postscript.

9. Here we might recall Nietzsche's notion of "purposeful forgetting," especially in the last sections of *Gay Science*. Serious literary Poundians have of course gone much further in reconstructing his intentions and meanings from forgotten or unwritten works than have philosophical interpreters of Nietzsche. Derrida seeks to repair this sort of discrepancy by interpreting even the smallest fragments of Nietzsche's "mad" ramblings by borrowing and ironizing a variety of critical tools. In *Spurs*—with a wink at Foucault and the other editors of the 1965 French translation of Nietzsche's complete works—he mocks genealogical seriousness by elaborating various interpretations of the Nietzschean fragment "I have forgotten my umbrella." He concludes that Heidegger's "Question of Being" and his reconstruction of *Will to Power* are involved in an ongoing forgetfulness: "Although I have read and quoted it, I no longer recalled this text of Heidegger from *Zur Seinsfrage*: 'In the culminating phase of nihilism it seems that there is no such thing as the *Being of being*,' that there is nothing to it (in the sense of *nihil negativum*). Being remains absent in a singular way. It veils itself. It remains in a veiled concealment (*Verbogenheit*) which veils itself in itself. In such a movement, then, consists the essence of forgetting which is experienced in the way the Greeks experienced it." Mocking the Heideggerian and subsequent readings of Nietzsche, Derrida reinserts the umbrella joke at the end of his second postscript. Pound's Kulchur as a forgetting that must forever frustrate academic critics is nearly echoed here: "Thus, in a thousand ways, has the 'Forgetting of Being' (*Seinvergessenheit*) been represented as if Being (figuratively speaking) were the umbrella that some philosophy professor, in his distraction, left somewhere." Jacques Derrida, *Spurs: Nietzsche's Styles,* trans. Barbara Harlow, preface by Stefano Agosti (Chicago: University of Chicago Press, 1971), pp. 141–43.

10. See Joseph N. Riddel, "'Neo-Nietzschean Clatter': Speculation and the Modernist Poetic Image," *Boundary 2* IX, no. 3, and X, no. 1 (Spring/Fall 1981): 209–39. Reading Pound's notion of "spermatazoic" thought against Nietzsche's *Will to Power,* section 805, Riddel suggests its implications for language and poetic tradition by uncovering a pun, about which I will have more to say below: "Pound's metaphor of the seminal ('semen,' *seme*) 'brain' is a figure for the 'poetic reserve' he elsewhere calls 'tradition' of the always already play of images" (p. 221).

11. Wyndham Lewis, critic, artist, novelist-philosopher, even more the

polymath than Pound, redefined the Shakespearean hero along Machiavellian-Nietzschean-Marlovian-Senecan lines in *The Lion and the Fox: The Role of the Hero in the Plays of Shakespeare* (London: Grant Richards, 1927). Eliot found this book and Lewis's place in Shakespeare criticism worthy of comment in his famous essay, "Shakespeare and the Stoicism of Seneca." Lewis's revaluation of violence and sensuality can also be *seen* in his Vorticist series of drawings on Shakespeare themes, the "Timon of Athens" suite being the most familiar.

12. Gourmont's obvious debt to Nietzsche in this regard was recognized by Richard Aldington in *Remy de Gourmont: A Modern Man of Letters* (Seattle: University of Washington, 1928), p. 8 *passim*. Almost forgotten today, Gourmont was acknowledged by his Anglo-American contemporaries to be the exemplary modern humanist. Similarly, Pound virtually bypasses Nietzsche in the two major essays he published on his favorite Frenchman, "Remy de Gourmont: A Distinction," *Instigations* (New York: Boni & Liveright, 1920) and "Mr. Aldington's Views on Gourmont," a short version of which appeared in *Dial* LXXXVI (January 1929): [68]–71. For a traditional history of Gourmont's influence on Pound, see Richard Sieburth, *Instigations: Ezra Pound and Remy de Gourmont* (Cambridge, Mass.: Harvard University Press, 1978), an attempt rigorously to account for the contradictions within Pound's Imagist and other, early aesthetics.

13. Nietzsche's influence on modernism was consciously denied, for reasons of politics—more than to avoid influential (or Bloomian) anxiety. Prominent Anglo-American artists and critics who had previously acknowledged Nietzsche's importance grew silent concerning him during and after World War I. Even obscure provincial journals asked questions such as "Did Nietzsche Cause the War?" *Educational Review* (Easton, Pennsylvania) V, no. 48 (1914): 353–57.

14. Paul de Man, *Allegories of Reading* (New Haven: Yale University Press, 1979), p. 118.

15. Paul de Man, "A Letter," *Critical Inquiry* 8, no. 3 (1982): 512.

16. Derrida, *Spurs*, p. 101, goes so far as to adopt Nietzsche's manner of undoing philosophical seriousness thus: "Somewhere parody always presupposes a naivety [*sic*] withdrawing into a consciousness, a vertiginous non-mastery."

17. *Ibid.*, p. 99. Of course, Heidegger is neither the first nor the last philosopher to puzzle over Nietzsche's significance to philosophy. Outside of Derrida and his followers, Arthur Danto, still a "serious philosopher" by his own accounting, represents the most sustained contemporary attempt to master and yet remain faithful to Nietzsche. See Arthur C. Danto, *Nietzsche as Philosopher: An Original Study* (New York: Columbia University Press, 1965; Morningside Books, 1980). That work is both a last-ditch effort to rephilosophize Nietzsche and clear proof that the virulence of the deconstructive project spread as far as the Harvard philosophy department. Is nothing sacred? Is such playfulness inevitable? Danto would seem, at least tacitly, to proclaim as much. For example, in his 1980 preface, before his infection by Derrida, he adopts such bad (i.e., deconstructive) habits as puns, false etymologies, and the apparently narcissistic reflection on his own signature. His epigraph reads: "*arthurdantist*, n., One who straightens

the teeth of exotic dogmas. 'Little Friedrich used to say the most wonderful things before we took him to the arthurdantist'—Frau Nietzsche.' Daniel Dennett and Karel Lambert, *The Philosophical Lexicon.*" In his belated preface, he identifies his own version of an appropriately philosophical Nietzsche: "There were lots of Nietzsches. He belongs to the histories of philosophy, opera, ideology, and hermeneutics, to the chronicles of loneliness, madness, and sexual torment. Mine is the philosopher" (p. 10).

18. Derrida, *Spurs,* pp. 72–75. This unplaceable, uninterpretable, or "unreadable" fragment, excluded from the definitive editions of Nietzsche's collected works, easily instigates an elaborate interpretation of repressed sexuality of Nietzsche's pointed reading/writing style and his inflated reading economy.

19. Notwithstanding Pound's insistence upon concision, DICHTEN= CONDENSARE (which translates into the ungrammatical "poeticizing means condensing"), *The Cantos* continues to grow posthumously, generating voluminous commentaries that seem to become part of that poem. Efforts to define the limits of Pound's works weight the reading economy on the side of interpretation. And even though the Pound industry lacks a deconstructive gadfly, its members continually come up against fundamental questions of literary criticism, such as the status of Pound's own disturbed scribblings and the genre of some of the fragments which now occupy files labeled "Collected Prose: Unpublished and Unsorted" but could as easily be drafts for the later Cantos.

20. In his characteristically ambivalent reaction to Nietzsche's pursuit of an aesthetic, Heidegger asserts that "Nietzsche's meditation on art is an 'aesthetics' because it examines the state of creation and enjoyment: It is the 'extreme' aesthetics inasmuch as that state is pursued to the farthest perimeter of the bodily state as such, to what is farthest removed from the spirit, from the spirituality of what is created" (N, 129). This seems close to the Nietzschean affirmation of modern decadence in the historicity of its own aesthetic: "Aesthetics is tied indissolubly to these biological presuppositions: there is an aesthetics of *decadence*, and there is a *classical* aesthetics—the 'beautiful in itself' (*Schönes-an-sich*) is a figment of the imagination, like all of idealism" (CW, 190).

21. This distinction is fundamental to the continuing argument about poetic influence in, for example, the works of Harold Bloom, especially in his efforts to deflect the full impact of Derrida's interrogation of origins and his notion of "dissemination." Bloom's efforts could be said to describe or reinscribe Heidegger's *Überwindung* against the Derridian-Nietzschean *Verwindung*, the latter concept/word/metaphor affirming the radical or originary play of language against Bloom's theological notion of a fixable Origin and a defined tropology of repression and transumption.

22. Building and inhabiting philosophical structures, if not "houses of fiction," are Heidegger's favorite metaphors (borrowed from, among other places, Kant's Third Critique) for philosophical system building. For repetitions and elaborations of architectonic figures, metaphors for grounded philosophical language or "metaphors for metaphor," see "Building Dwelling Thinking" in PLT.

23. If Nietzsche did not use the phrase "dithyrambic critic" to describe himself, Wyndham Lewis, whose Nietzscheanism is too simply dismissed as proto-fascist rather than analyzed as an interesting *rhetorical* analysis, elaborates this notion in *The Diabolical Principle and the Dithyrambic Spectator* (London: Chatto & Windus, 1931).

24. Nietzsche's physiological metaphors for the origin of philosophy in poetry and physical sensation and for the conceptual process often seem to echo Wagner's birth images for artistic creativity (these can also be found in Schopenhauer's nearly pornographic accounts of the relations between music and philosophy). These cross-discursive tropologies also occur in the "Translator's Postscript" to *The Natural Philosophy of Love,* where interpretation becomes a sexual act. Let me here cite Wagner's physiological and onto-theo-logical account of the birth of *Gesamtkunstwerk* out of *Zukunftsmusik:* "As man by love sinks his whole nature in that of a woman, in order to pass over through her into a third being, the child—and yet finds himself again in all the loving trinity, though in this a widened, filled, and finished whole . . . on the path of genuine love and by sinking of itself within the kindred in the perfect love and by sinking of itself within the kindred arts if [music] returns upon itself and finds the guerdon of its love in the perfect work of art to which it knows itself expanded." *Wagner on Music and Art: A Compendium of Wagner's Prose Works,* trans. H. Ashton Ellis, ed. Albert Goldman and Evert Spinchorn (New York: Da Capo, 1981), p. 121.

25. Nietzsche frequently mocks Wagner's adoption of Schopenhauer as a corruption (a failed purification) of art by metaphysics. In *The Case of Wagner,* he jokingly exposes philosophy's double repression of sexuality and Christianity, thus: "Brunhilde was initially supposed to take her farewell with a song in honor of free love, putting off the world with the hope for a socialist utopia in which 'all turns out well'—but now gets something else to do. She has to study Schopenhauer first. . . . *Wagner was redeemed.* In all seriousness, this *was* a redemption. The benefit Schopenhauer conferred on Wagner is immeasurable. Only the *philosopher of decadence* gave the artist of decadence—himself" (CW, 164). That passage abbreviates Wagner's biography from socialism (in the Paris trenches of 1848) to political conservatism and the theologization of art; it attacks the notion of authorial identity and control, if it does not mock as illicit the relations between Romantic art and metaphysics.

26. While for Nietzsche "decadence" was not a wholly negative term, he sees Wagner's masterpiece, *The Ring,* as a decline from Goethe's affirmative Romanticism: "Goethe once asked himself what danger threatened all romantics: the falsity of romanticism: his answer was: 'suffocating on the rumination of moral and religious absurdities. In brief, *Parsifal*" (CW, 162). Protests to the contrary notwithstanding, Nietzsche had to guard himself against such ruminations. Much in the same way, Pound seemed to fight a rear-guard action against theology and "Reverend Eliot," or that modernism which grounds traditional humanism in the various dogmas of classicism, Christianity, and the like.

27. Nietzsche contrasts the Renaissance Italian quality of *virtù,* strength, re-sponsibility, and manly nobility to Christian virtue in *Genealogy of Morals* and elsewhere. For example, a footnote in *The Case of Wagner,* which Kaufmann's footnote tells us contains the only three footnotes Nietzsche ever cared to write, calls attention to *Genealogy of Morals* in this regard. One is familiar with this same distinction in the histories Pound borrowed from Sigismundo Malatesta, Gourmont, Jacob Burckhardt, and a whole host of vaguely Nietzschean sources, including Wyndham Lewis.

28. One is again reminded of Heidegger's trope of philosophical system build-ing and most philosophers' preference for classical architectural (and "arche-tex-tual") metaphors. Derrida follows this trope from Kant's Third Critique through Heidegger's "Origin of the Work of Art" to Meyer Schapiro's "Some Reflections on a Still Life"; see *Vérité en Peinture* (Paris: Flammarion, 1978).

29. Theodor Adorno, *In Search of Wagner,* trans. Rodney Livingstone (Lon-don: New Left Books, 1981), p. 48.

30. Here and throughout, I mean privilege in two senses, but especially in the technical sense of assigning a special value. First, *privilege* as or from an ad-jective, recalls the privileged reader of Eliot and the New Critics, who acknowl-edged that a privileged few were—by instinct and training, if not by birth—more capable of understanding literature than others. Second, *to privilege,* as a verb (which might more conservatively but less euphonically be used gerundively, "to grant a privilege to"), has the more technical meaning of making a proper name, work, concept, or style represent and justify a whole ouvre, discourse, age, or culture. In the crucial example of Nietzsche reprivileging the neglected and re-jected fragments out of which Wagner was forced to build his huge productions, the very act of privileging is questioned or made ironic. Not only do the acci-dental and unintended parts of Wagner's art jar with his operas, but, when they are picked out and highlighted, they subvert from within the composer's as well as the audience's dream of a complete art. In sum, Nietzsche exposes Wagner's fragmented masterpieces only to bare the workings of the culture industry, that series of vested relationships that determines taste and value. I have deliberately avoided *valorize* in this context. As a term borrowed from Portuguese in the late Renaissance, valorize, as Pound properly uses it in his economic writings, means the assignment of a fixed value without regard to market changes. This is quite distinct from the arbitrary and differential assignment of privilege, even if priv-ilege can itself become the measure or guarantor of power and value.

31. Despite his position as one of the major exponents of a certain modernist style (Imagism, as a precision of language exclusive of rhetoric), potential vol-umes of Pound's critical prose lie uncatalogued, unsorted, and open to reordering by every new reader of the folders at the Beinecke, for example. The folders of "Collected Prose: Unpublished and Unsorted" were apparently assembled by Pound in about 1931 when Louis Zukofsky was supervising publication of the *Prole-gomena Series,* which was to include all Pound's critical prose. The project ceased with the bankruptcy of the publisher. The only volume issued was a reprint of two earlier essays, *How to Read* with *The Spirit of Romance* (Le Beausset, France:

To Publishers, 1932). In 1959, when he again reviewed the material he had given to Yale, Pound partially resorted and randomly annotated "Collected Prose" and other files. Since Pound seems himself to have misidentified several of the typescripts and notes, and since there is every possibility that this material will be moved again for cataloging and easier access, I will refer to all unpublished prose simply as Beinecke prose, with file and page number only when it is clear that sheets of prose will be kept together.

32. Pound deliberately opens his criticism to an uncontainable tropology. In what an earlier generation of formalists (New Critics) might excuse as the poet's sensibility or license, he privileges metaphors over the precise and stable definitions of philosophy and serious criticism: "I ask the reader to regard what follows not as dogma, but as a metaphor which I find convenient to express certain relations," "I Gather the Limbs of Osiris," *The New Age* X, no. 10 (11 January 1912): 224. Indeed, such "methods" as "The Method of Luminous Detail" and "Paideuma" are metaphors—if, as he says, interpretive metaphors—first and only secondarily methods.

33. Pound's reading *practice* in his never completed or conclusive essays, notes, and fragments fails to make a *theory*. Yet he argues implicitly against clear distinctions and hierarchization of reading or interpretation and writing. Traditionally, reading is the mastery of the play or undecidability of writing, which is privileged in "poetic" but anathema to "philosophic" writings. The deliberate breakdown of this hierarchy and the distinction on which it rests are part of Nietzschean and, I would claim, Poundian rigor. Jacques Derrida, even more radically than Paul de Man, describes this (his own) rigor, which has been considered as "defective" and aberrant as Pound's. Here the immediate text under analysis is Plato's *Phaedrus*, but the musical metaphors might lead us back through Nietzsche to Wagner: "The hypothesis of a rigorous, sure, and subtle form is naturally more fertile. It discovers new chords, new concordances; it surprises them in minutely fashioned counterpoint, within a more secret organization of themes, of names, of words. It unties a whole *sumploke* patiently interlacing the arguments. What is magisterial about the demonstration affirms itself and effaces itself at once, with suppleness, irony, and discretion," Jacques Derrida, *Disseminations*, trans. Barbara Johnson (New Haven: Yale University Press, 1981), p. 67.

34. Hugh Kenner, *The Poetry of Ezra Pound* (London: Faber & Faber, 1951); especially pp. 49–55. Hereafter noted as PP in my text.

35. Hugh Kenner, *The Pound Era* (Berkeley: University of California Press, 1971). Hereafter cited as PE in my text.

36. Kenner detects Pound's Imagism, a program and quite literally a metaphor of metaphor which Kenner reads as a return to "Aristotelian" clarity of definition, behind Eliot's poetry. This seems safe enough, but Kenner refuses to interrogate the notion of Imagism in terms of its subversion of logic. Instead, he contents himself with the not inconsiderable trope of making Pound precursor to Eliot's most orthodox works. For another, quite different instance of ventriloquism, see note 15 of Chapter Two.

37. Kenner, PE, 49.

Chapter Two

1. "Placidity," an unpublished essay in the Harriet Monroe *Poetry* Collection, University of Chicago, p. 2.

2. Beinecke prose.

3. Such simple interrogative sentences, inviting equally simple answers, seem as inadequate to Pound's reading method as Alphonso Lingis has found them to Nietzsche's Will to Power, when he says: "We would put to Nietzsche the familiar form of the philosophical question. It asks after the essence of the Will to Power. The philosophical question 'what is . . . ?' is answered by supplying the quiddity, the essence. Philosophical thought is a questioning of appearances, an investigation of their essence, their organizing structure, their telos, their meaning. . . . Philosophical interrogation of the world is a reading of the world, an assumption of the succession of sensorial images as signs of intelligible essences. Nietzsche refuses this reading of the world." Alfonso Lingis, "The Will to Power," in *The New Nietzsche,* p. 37.

4. Emerson's "The American Scholar" advertises a similar notion of interpretation, attributing creativity and mystical—if not metaphysical—"luminosity" to the traditionally devalued act of reading: "One must be an inventor to read well. . . . There is creative reading as well as creative writing. When the mind is braced by labor and invention, the page of whatever book we read becomes luminous with manifold allusion. Every sentence is doubly significant, and the sense of our author is as broad as the world." *The Complete Works of Ralph Waldo Emerson,* 12 vols., intro. and notes by Edward Waldo Emerson (Boston: Houghton Mifflin, 1903–1906), vol. I, pp. 42–43. Hereinafter cited in text by volume and page number.

5. As noted earlier, the victory of Eliot's organic tradition and his notion of a natural language (much satirized by Pound in, say, Canto LXVI), is confirmed when Kenner legitimizes Pound by associating him with Eliot, the great legitimizer of what would become the New Critical canon. On the other hand, Terry Eagleton sees Eliot's deployment of the timeless yet natural ("organic") model of literature as a brilliant stroke of political reaction and cultural conservatism— his reading of the unity beneath the apparent fragmentation of *The Waste Land* is quite convincing. Eagleton goes so far as to credit Eliot with a culturally rightist revolution which he claims grew, as though organically, from the agrarian South (not St. Louis) which Eliot visited only late in life for lectures—though Eliot was adopted by the New Critics, most of whom were in Tennessee, not Virginia. Eagleton's own British "regionalism" aside, he offers this provocative reading of the power of Eliot's mild-mannered "anti-ideology": "Spiritually disinherited like James by industrial capitalist America, able later to discover in America the 'blood', breeding and 'organic' regionalism he valued only in such phenomena as the right-wing neo-agrarian movement in Virginia, Eliot came to Europe with the historic mission of redefining the organic unity of its cultural traditions, and reinserting a culturally provincial England into that totality. He was indeed to become the focal-point of the organic consciousness of the 'European mind', that

rich, unruptured entity mystically inherent in its complex simultaneity in every artist nourished by it. English literary culture, still in the grip of ideologically exhausted forms of liberal humanism and late Romanticism, was to be radically reconstructed into a classicism which would eradicate the last vestiges of 'Whiggism' (protestantism, liberalism, Romanticism, humanism). It would do so in the name of a higher, corporate ideological formulation, defined by the surrender of 'personality' to order, reason, authority and tradition," *Criticism and Ideology* (London: Verso, 1978), p. 146. See also his "The Rise of English," *Literary Theory, An Introduction* (Minneapolis: University of Minnesota Press, 1983), pp. 17–53.

6. Pound was more in reaction against than in support of philology and the organic/genealogical theory of language Kenner suggests; indeed, in tenuous alliance with Leavis, Eliot, and Hulme, he wished, even as late as *ABC of Reading*, to wrest literature from the German and Oxford philologists. Just so, Pound finds in translation the greatest potential for creative and wayward interpretation, for disrupting cognition and mocking the very notion of translatability that rests on perfect cognates. See especially his essay on the Elizabethan Classicists. For a similarly ambivalent account of translation, compare Walter Benjamin's essays "The Task of the Translator," *Illuminations,* ed. Hannah Arendt (New York: Shocken, 1969), and "On Language as Such and One Language of Man," *Reflections,* ed. Peter Demets (New York: Harcourt Brace Jovanovich, 1978).

7. Paul de Man, "Introduction," *Towards an Aesthetic of Reception,* by Hans Robert Jauss, trans. Timothy Bahti (Minneapolis: University of Minnesota Press, 1982), pp. xv–xvi.

8. Jacques Derrida makes a similar point with regard to metaphysics, which he finds to be grounded in "rhetoric," rather than in (Heidegger's) Being or even beings, that is, in tropes rather than in an ontological referent: "Our hypothesis then is that neither a rhetoric of philosophy nor a metaphilosophy is to the point. Why should we not start with rhetoric as such?" "White Mythology: Metaphor in the Text of Philosophy," *New Literary History* IV, no. 1 (Autumn 1974): 31. Hereinafter cited as WM.

9. A curious passage in *Guide to Kulchur* links the names of Heraclitus and Aristotle, much to the detriment of the latter, whose "logic," Pound says, has declined from Heraclitus's "rhetoric": "The profound flippancy of a decadence, when men don't know when to stop talking and have ceased to respect the unknown! Very well! Arry teases Heraclitus for vehement assertion, but is that assertion any worse than silly logic chopping on the same matter?" (GK, 340). Elsewhere, Pound associates Heraclitus with the axiom "everything flows," in the sense that I have opposed "Heraclitan" to Aristotle's categories in an analysis developed in Chapter Five.

10. Derrida mentions Pound and Mallarmé in passing, suggesting that their "graphic poetics" marks a significant break in the Western category "Being," which has always privileged "speech" over "writing." See *Of Grammatology,* p. 92. Joseph Riddel uses Derrida's suggestion as an "instigation" for "'Neo-Nietzschean Clatter.'"

11. While speculation on the source and depth of Pound's knowledge of formal or systematic rhetoric yields no final answers, one might note that one of the classical rhetoric handbooks, Cicero's *Rhetorica ad Herennium,* defines metaphor in terms similar to those Pound attributes to Aristotle. *Translatio,* the Latin translation of "metaphor," refers directly to movement and "carrying over." Cicero writes: "Metaphor occurs when a word applying to one thing is transferred to another, because the similarity seems to justify transference," *Ad Herennium* (Loeb Classics) IV, p. xxxiv. This is more a reading than a simple rendering of Aristotle, because it elevates exchange and transformation from examples to defining principles. As is obvious, this plays doubly on accidents of language; *translatio* being an even more insistent pun than *Übertragung* and *Übersetzen.*

12. Using Buckminster Fuller's figure of cultural nodes or "knots," Kenner links East and West (Emerson, Hegel, Confucius, Buddha, and other traditional philosophical and religious figures) as the proper sources and analogues of Fenollosa's readings of Chinese poetry and the written character. See Kenner, *The Pound Era,* (Berkeley: University of California Press, 1971), especially pp. 192–222. One might say that here again he grants Pound System, but at the expense of that poet's own *antisystematic* (if not anti-Hegelian, then antiphilosophical and anti-Christian) reading procedure.

13. *The Poetics* develops the distributive function of metaphors of proportion with regard to the sun, that most privileged of Western tropes, which cannot be seen or signified directly: "It may be that some of the terms thus related have no special name of their own, but for all that they will be metaphorically described in just the same way. Thus to cast forth seed-corn is called 'sowing'; but to cast forth its flame, as said of the sun, has no special name. This nameless act (B), however, stands in just the same relation to its object, sunlight (A), as sowing (D) to the seed-corn (C). Hence the expression in the poet 'sowing a god-created flame' (D & A)." *The Poetics* 1457b, 25–30; cited in Derrida, WM, 43–44. Pound's notion of "metaphor" and Ray's Ideogram, which has the sun as its ultimate referent, both imply this notion of metaphor.

14. Jacques Derrida, "The *Retrait* of Metaphor," *Enclitic* 2, no. 2 (Fall 1978):6. Hereafter noted as RM in my text. The passage immediately following the one cited more directly remarks the inescapability and uncontainability of figurative language in connection with the impossibility of a *meta-metaphorics:* "The drama, for this is a drama, is that even if I had decided to no longer speak metaphorically about metaphor, I would not achieve it, it would continue to go on without me, in order to make me speak, to ventriloquize me, metaphorize me. . . . Any statement concerning anything that happens, metaphor included, will be produced *not without* metaphor. There will not have been a meta-metaphorics consistent enough to dominate all its statements" (RM, 8).

15. Clipping a passage from *Littré,* the French OED, Derrida presents without comment—but as a commentary on philosophy's habitual dwelling upon borrowed terminology—the following definition of catachresis: "Trope par lequel un mot détourné de son sens propre est accepté dans le langage commun pour signifier une autre chose qui a quelque analogie avec l'objét qu'il exprimait

d'abord; par exemple, une langue, parce que la langue est le principal organe de la parole articulée; une glace . . . une feuille de papier," *Glas* (Paris: Denoël/ Gonthier, 1981), p. 2. Such a figure describes the graphic displacement, the superpositioning of philosophic upon poetic texts, in Derrida's later works, especially *Glas,* in which allusions ring across three languages and interrupt each other in the text's three columns. Curiously, Derrida simply found ready-made an example of the way philosophical language is abused by the common "tongue": language—or "language"—itself is a catechresis. For a consideration of philosophy as a closed system against philosophy as a continuing and revisionary dialogue, see Richard Rorty, *Philosophy and the Mirror of Nature* (Princeton, N.J.: Princeton University Press, 1979).

16. "And The True History," Beinecke prose.

17. I refer to de Man's sometimes deliberately muted distinction between irony as a figure and as an essential condition of language.

18. Not unlike Pound, Walter Benjamin subscribed to the notion of a horizon or atmosphere of poetic meaning that could not be translated. In "The Task of the Translator," Benjamin outlines two independent criteria of translatability: informational accuracy, which can be achieved most fully prosaically and on the lexical level; and a fluxional meaning that might capture the vital significance that dances between lines of poetry, for which an "interlinear" translation of the Scriptures is his ideal.

19. The fantasy of the completion of the world on the model of the sentence, the complete sentence as proof of a coherent universe—or God—is at least as old as Augustine. Calling upon his pre-Christian experience as a rhetorician, Augustine elaborates the sentence as proof that God is the resolution of time in eternity, in *Confessions,* Book II, chaps. 25–28. Fenollosa could have taken his suggestion from any number of intermediate sources, especially the Romantics; Coleridge and Keats both mention the world as a completable sentence.

20. Somewhat belatedly, then, Pound positions himself on the methodological chiasmus of eighteenth- and nineteenth-century sciences that recent—mostly French and neo-Nietzschean—philosophers have found so suggestive. For example, risking the historicist's oversimplification that Michel Foucault condemns, I would cite his reading, in *The Order of Things (Les Mots et les choses),* of the discursive change from "natural history" to "biology"—from analyses of living things to a concept of "life," which includes temporality and natural forces. The following uses the library as an emblem of the infusion of historicity into classificatory systems, bringing the "scientific revolution" into the realm of letters: "The ever more complete preservation of what was written, the establishment of archives, then of filing systems for them, the reorganization of libraries, the drawing up of catalogues, indexes, and inventories. . . . It is this classified time, that the historians of the nineteenth century were to undertake, the creation of a history that could at last be 'true'—in other words, liberated from Classical rationality, from its ordering and theodicy: a history restored to the irruptive violence of time." *The Order of Things: An Archaeology of the Human Sciences* (New York: Vintage Books, 1973), p. 132.

21. *Leo Frobenius 1873–1973: An Anthology,* ed. Eike Haberland (Wiesbaden: Franz Steiner, 1973), p. 6.

22. Sarah Kofman, in "Metaphor, Symbol, and Metamorphosis," cites Nietzsche's reading of Aristotelian metaphor as an instance of his critique of "essences": "For Aristotle, the concept is prior in relation to the metaphor . . . as the passage from one logical place to another, from a 'proper' to a figurative place. The Aristotelian definition of metaphor as such cannot be retained by Nietzsche, for it rests on a division of the world into clearly defined genera and species, which correspond to essences" (NN, 207).

23. Nietzsche makes clear that perception, like judgment, is a highly organized system of selection and interpretation, subject, as is belief, to Will, History, and more general cultural forces. See especially WP, section 505.

24. Against Nietzsche, Heidegger claimed that—by virtue of Will's status as an ontological category or even another name for Being (*Sein*)—Nietzsche was the last of the great German metaphysicians. To do so, he had to redeem Nietzsche from the self-proclaimed category of "poetic revolutionary." Curiously enough, William Carlos Williams subjected Pound to a similarly willful misreading. In a gesture he would later recant in light of Pound's later poetry, he reinscribed his American cohort into the evolving English and Eliotic tradition with such phrases as "men content with the connotations of their masters" ("Prologue to Kora in Hell") and "traditionalists of plagiarism" (*Spring and All*). See Chapter Four herein.

25. Along these lines, one would want to consider Bataille, Blanchot, Deleuze, Adorno, Benjamin, de Man: on the whole, hardly a school or even a distentanglable network of influences; as such this grouping repeats the vaguely Nietzschean "air" surrounding Pound. One thing is clear, however: as in Pound's time, Nietzsche, notwithstanding interested readings and translations, is again an important force.

26. Herbert N. Schneidau begins to uncover the disruptive potential of the formula when, in commenting on Pound's choice of the Bible, Ovid's *Metamorphoses,* and *The Canterbury Tales* as examples of DICHTEN= CONDENSARE at work, he focuses on fragments rather than the ideal of "the whole work": "Pound apparently believes that if a poet really concentrates his work, the logical result is a compendium; a long unified tale can be produced only by spreading the effort. He assumes that the immortality of his illustrative works is due to such concentration. The list of examples might be extended. . . . The applicability to the *Cantos* is obvious." *Ezra Pound: The Image and the Real* (Baton Rouge: Louisiana State University Press, 1969), p. 139.

27. In a footnote to Pound's repetition of DICHTEN=CONDENSARE, in "A Visiting Card, 1942," Noel Stock treats Pound's erroneous etymology. However, rather than fully exploiting the complexities of *Dichten,* he merely suggests that it is not etymologically related to "to tighten." Again the focus is on condensing, though Stock emphasizes the most mystical aspects of *Dichten,* by stressing poetic intuition over philological rigor: "The two words dichten (to draw in, tighten) and dichten (to write poetry) appear to have different roots. Whether the poet's

intuition has seen deeper than philology, remains to be determined" (IM, 66). A contemporary reader like Derrida might turn that etymological accident into an "undecidable" in order to indicate how the notion of "condensation" also contains its opposite, "dissemination," or to argue that the two different words in an uncanny way translate the "effects" of metaphor.

28. In an essay concerning Nietzsche's disturbance of the philosophical *Darstellung*, and the treatment of these by Heidegger and Derrida, Eugenio Donato finds "The Saying of Anaximander," the Heidegger essay I have quoted, particularly suggestive. See "The Idioms of the Text: Notes on the Language of Philosophy and the Fictions of Literature," *Glyph 2* (1977), pp. 1–13.

29. "Phanopoeia," *Collected Prose*, Beinecke prose, p. 1. This two-page note, which concludes with the metaphor of "The rose in steel dust," appears to be from the 1930s, when Pound reviewed his critical essays for reissue. "The rose in steel dust" also appears, among other places, in Canto LXXIV, nor is it unrelated to the figure of iron filings moved by a magnet held under glass in "I Gather the Limbs of Osiris." In all these instances this figure of "patterned energy" shows Pound's desire to have a system at once "natural," that is, organic, whole, and without human interference, and *totalitarian*—by which he meant not only controlling but also the product of a total art or *technē*.

Chapter Three

1. T. S. Eliot, "Ezra Pound: His Metric and Poetry (1917)," *To Criticize the Critic and Other Writings* (New York: Farrar, Strauss & Giroux, 1965), p. 163.
Quotations from other books by Eliot will be cited in the text as follows:

ASG *After Strange Gods: A Primer of Modern Heresy* (New York: Harcourt, Brace, 1933).

DC *Notes Toward the Definition of Culture* (New York: Harcourt, Brace, 1949).

SE *T. S. Eliot: Selected Essays, New Edition* (New York: Harcourt, Brace & World, 1960).

UP *The Use of Poetry and the Use of Criticism* (Cambridge, Mass.: Harvard University Press, 1933).

2. In one particularly suggestive letter (I have been able to consult only an unauthorized transcription at the Beinecke of an original allegedly held at Harvard's Houghton Library), Pound seems to anticipate and yet revise Eliot's "Tradition and the Individual Talent" (first published in the *Egoist*, 1919). While advocating the use of traditional texts, Pound lays stress on revolutionizing change over accommodation. At the moment, I can only paraphrase part of this letter to Henry Ware Eliot, Eliot's father, whom Pound somewhat unsuccessfully hoped to reassure about the efficacy of his son's choice of poetry as an occupation: "The arts, as the sciences, progress by infinitesimal stages, that each inventor does little more than make slight, but revolutionizing change, alteration in the work of his predecessors." The Beinecke typescript gives the postmark date of that letter as 28 June 1915. I would want to consult this and other now unavailable

early letters before attempting to argue priority and coherence of Pound's formulation of an iconoclastic tradition. Further, Eliot's own earlier elaboration of the familiar thesis of "Tradition and the Individual Talent," the essay "Reflections on Contemporary Poetry," *Egoist* (July 1919), is more disruptive of tradition. Harold Bloom cites this essay as an instance of Eliot's revisionary tendencies (which were, as we have seen, already revised by late 1919). See "Ratios," *The Breaking of the Vessels* (Chicago: University of Chicago Press, 1982), pp. 18–19.

3. "Eeldrop and Appleplex," *Little Review* IV, no. 1 (May 1917): 7.

4. *Ibid.*, p. 8.

5. Pound to Harriet Monroe, 22 October 1912 (Harriet Monroe *Poetry* Collection, University of Chicago). In an unpublished section of what was originally a twenty-one-page letter, Pound adopts the role (or "mask") of scourge or gadfly of complacency: "Print me with an asbestos border & deplore my opinions in a footnote or in an editorial. . . . *I know perfectly well that my criticism seems like the speech of a conceited fool. I accept this burden* along with a few other loads" (emphasis added).

6. Pound to Harriet Monroe, received 12 July 1913 (*Poetry* Collection, Chicago). In this early statement of his journalistic ideals, Pound makes clear his plan simply to make available reading material to the American public, particularly those writings rejected by establishment literary organs: "My conception of our function is that we should print, as much as possible, what is significant. Stuff that other periodicals can't or won't handle. Stuff that is interesting as ART. . . . Our second function is to develop the market." These goals and their allusion to the market economy of publication is consistent with the purpose Pound expressed for little magazines throughout his career.

7. F. R. Leavis, *How to Teach Reading: A Primer for Ezra Pound* (Cambridge: Gordon Frazer, Minority Press, 1932), pp. 18–19. This forty-nine-page pamphlet was written in reply to "How to Read," which Leavis reads both as a textbook and as an attack on Eliot. Ironically, Leavis values Pound for the very things, iconoclasm and unconsciousness, that as moral critic he must censure. For example, his "Tribute and Valediction to Pound" skirts the pedagogical issue of Pound's radical questioning, to which Leavis feels a sort of guilty attraction: "He may deny that he raised them, but that he did raise them is the reason for treating his pamphlet seriously, and *that he did not know he raised them is perhaps the radical criticism of Mr. Pound*" (p. 49; emphasis added). Leavis, like Eliot, had good reason to admire Pound's enthusiasm, even as he hoped to blunt the unreconstructed modernism of that other American reader/writer. Leavis wanted to substitute for the old order a renewed literary order, while Pound seemed bent on reordering or disordering the means by which literature was institutionalized. Leavis's *Scrutiny*, though outside the scope of the present study, bears consideration in this regard.

8. Pound to Eliot [1934], Beinecke.

9. "Table of Contents," Beinecke prose. This table, which bears little relation to the contents of the file, was apparently intended for Louis Zukofsky's projected

edition of Pound's complete prose. The note cited and several other introductory fragments to which I refer offer self-criticisms, not always simple corrections or additions, to Pound's earlier critical essays.

10. Pound to Eliot, 15 September 1959, Beinecke. This same letter refers to *After Strange Gods* as a shocking example of Eliot's conservatism, his reaction against experimental art and criticism. Pound had already made himself a counterculture hero on the model of Malatesta.

11. Beinecke prose. Here Pound's *ungrammaticalness*—his composition slippages—apparently result from haste and colloquialism; elsewhere he launches attacks on Western grammar, which is necessary to (onto-theo-logical) Western thought. For example, in *The Chinese Written Character as a Medium for Poetry*, he and/or Fenollosa insist(s) that there are no pure nouns and verbs, thus no proper names or identity, in nature—or language *properly used*. This radicalization of the question "What is language?" is posed throughout Pound's reading text, but nowhere as clearly as in his fascination with the Chinese *ideogram*.

12. 15 January 1934, Beinecke.

13. Eliot to Pound, 25 January 1934, Beinecke.

14. While Pound clearly disapproved of the London literary establishment, his far-flung poetic and journalistic activities belie any general alienation or disaffection from *literature*—both as a series of texts and as a university discipline in the making. In a letter to Harriet Monroe, 13 December [1932] (Morton Zabel Papers, University of Chicago), Pound says: "I do not feel any lonelier now than I did in 1912; and the new generation of 1932 seems to me as promising as that of 1912."

15. Eliot to Pound, 25 January 1934, Beinecke.

16. Along these lines, though he complained bitterly of his unsuccessful attempts to publish *Jefferson and/or Mussolini*, he suggests that even his most dogmatic work was primarily intended to provoke controversy and stimulate the free exchange of ideas. In unpublished prefatory matter to that book, he says: "No typescript of mine has been read by so many people or brought me such interesting correspondence." "Jefferson and/or Mussolini" folders, Beinecke.

17. Eliot to Pound, 1 January 1959, Beinecke.

18. Pound to Eliot, 10 February 1935, Beinecke. Almost obsessively, Pound attends to details of presentation: orthography, subtitles, footnotes, page numbers, apparent typos, and so on. He often grounded his attacks of Eliot and others on these accidents customarily ignored in critical exchanges among the members of a discursive community. Pound's epistolary assaults suggest Jacques Derrida's more rigorous interrogations of such conventions in "Limited Inc abc," *Glyph* 2, pp. 162–254.

19. Beinecke.

20. Pound to Eliot, 5 February 1934, Beinecke.

21. *"Collected Prose,"* Beinecke, p. 5.

22. Tracing modernist fragmentation and the commodification of literature, Walter Benjamin analyzes the impact and ascendancy of literary sections—*feuilleton*—in French *journaux* in Baudelaire's Paris. Of such newspapers which, like Gour-

mont's *Mercure de France*, serialized novels and advertised or criticized literary currents, he says: "For a century and a half the literary life of the day had been centered around periodicals. Towards the end of the third decade of the century this began to change. The *feuilleton* provided a market for *belles-lettres* in the daily newspaper. The introduction of this cultural section summed up the changes which the July Revolution had brought to the press. . . . In 1824 there were 47,000 subscribers to newspapers in Paris; in 1836 there were 70,000 and in 1846, 200,000. In this rise Girardin's paper *La Presse* had played a decisive part. It had brought about three important innovations: the decrease of the subscription to forty francs, advertisements, and the serial novel" *Charles Baudelaire: A Lyric Poet in the Age of High Capitalism*, trans. Harry Zohn (London: New Left Books, 1973), p. 27.

23. "Commentary," *The Criterion* XII, no. 49 (July 1933): 647. While the focus on "value" and "economic emergency" echoes Pound's Social Credit concerns, Douglas Jerrold's *Symposium*—one of several neglected American Social Credit journals to which Pound, William Carlos Williams, and Robert McAlmon contributed literary and economic writings—is Eliot's immediate target.

24. For example, from 1931 to 1940, Pound wrote Iris Barry, then librarian of the New York Museum of Modern Art, proposing an art/literary/critical/anthropology journal. This venture, inspired in part by Frobenius's work on African art, never materialized, but the correspondence concerning it confirms Pound's opposition to Eliot's academicism and conservatism, an opposition that kept Pound marginal to the English belletristic tradition on which university literature curricula were (*are*) based. Iris Barry, 27 November 1933 (New York Museum of Modern Art Library), writes: "I would *not* use an article by Eliot that I have read, not interesting except for college professors." Her next letter continues this same theme: "I wouldn't use an essay by TSE if paid to do so, not in my street at all, not a university extension" (28 November 1933).

25. Pound to Eliot, [15] January 1935, Beinecke.

26. Pound to Eliot, 25 May 1935, Beinecke.

27. As previously remarked, Eliot had frequently to concern himself with Pound's inattention to copyright in the many quotations in his reading texts. Pound's most direct assertion of the intertextuality of his readings is in *Guide to Kulchur:* "There is no ownership in most of my statements and I cannot interrupt every sentence or paragraph to attribute authorship to each pair of words, especially as there is seldom an a priori claim to the phrase or the half phrase" (p. 60).

28. Import duties and copyright problems forced Pound to publish *The Exile* in Paris and Chicago, while he remained in Italy. This was not the multinationalism or transnationalism he sought, but it symbolizes his double foreignness—neither American nor Italian, always the self-appointed "exile" from (English) literature.

29. Such journals were numerous and popular. One such is *New Democracy: A Semi-Monthly Review of National Economics and the Arts*, an American version of *The New English Weekly*, which, with James Laughlin as literary editor,

published the writings of William Carlos Williams, Henry Miller, Kay Boyle, and e. e. cummings, as well as Pound's "Usury Cantos."

30. This eighteen-page untitled typescript accompanied a letter Pound sent to Eliot in hopes that he might redirect Possum's argument from (literary) orthodoxy to the consideration of the complex economics-literature-theology in the history of the Catholic church. Eliot's refusal to republish *After Strange Gods* forestalled Pound's attempted revisions. Indeed, Eliot's own disavowal of the text suggests that he found it already too unliterary—or "heretical," in its own terms.

31. Pound to Eliot, 25 February [1938], Beinecke. The approximate year assigned to this letter is erroneous, as Orage died 5 November 1934, and thus could not have made the editorial suggestions therein mentioned.

32. This might be mistaken for a continuation of Pound's old habit of blue-penciling the poems of his various epigoni—of whom Eliot was the most successful. But here, in these later revisions of Eliot, he seems quite deliberately to employ a sort of guerrilla editorializing upon his established (and establishment) friend.

33. As early as 1917 Pound identified American universities and journalism as his chief critical targets: "Just now there are two things to be eliminated 1. The German philological system in the universities 2. the old magazines." Pound to Harriet Monroe, 23 July 1917 (*Poetry* Collection, University of Chicago).

34. See Richard Sieburth, *Instigations*. This generally excellent source study fails to note the workings of "Dissociation" in Pound's later criticism. According to Sieburth, *The Natural Philosophy of Love* (1922) "closes Pound's Gourmontian decade" (p. 24). Sieburth follows Norman and others in marking a break between Pound's early French influences and his later German, Italian, and Fascist interests. But the neglected texts dealt with in this chapter put into question Pound's abandonment of his earliest critical formulations and borrowings, especially dissociation—even if the coherence (or is it merely a self-conscious, engineered continuity?) was read back into the uncollected prose by Pound himself in texts like *Guide to Kulchur*.

35. Glenn S. Burne, trans. and ed., *Remy de Gourmont: Selected Writings* (Ann Arbor: University of Michigan Press, 1966), p. 16. Hereinafter cited in text as SW.

36. Along these lines, as though again to dissociate himself from Eliot, Pound says: "We can also with advantage distinguish between the selective or critical faculty and the capacity for or habit of producing 'finished critical articles.' The faculty and the capacity are not mutually exclusive, neither are they invariably found together," "Collected Prose," Beinecke, p. 17.

37. In yet another folder of Beinecke's "Collected Prose," consisting of forty-one untitled leaves on Social Credit, Gesell, Douglas, and Mussolini quotations one of Pound's notes credits Douglas with the critical principle "increment of association": "A few men with a steam engine can sling a locomotive into the air, which 5000 savages can not."

38. This creative aspect of language and argumentation is already accounted

for in Gourmont's "dissociation" ("a new operation will disunite once again until the new ties, always fragile and equivocal, are formed" [note 35 above]). But Pound's belief in the freedom or indeterminacy of dissociation seems to fail him when he substitutes that metaphor for what he—perhaps erroneously—takes to be the univocal judgment "increment of association," which seems to suggest the kind of textual inflation or hoarded interest income that Pound elsewhere condemned by the name of usury and associated with the sclerotic tradition that halted the free exchanges of interpretations and figuration. Pound wanted more than one thing both ways.

Chapter Four

1. SP, 415. "Remy de Gourmont," *Fortnightly Review* (1915), a memorial essay written shortly after Gourmont's death.

2. LE, 357. This essay, "Remy de Gourmont: A Distinction," Pound's longest study of Gourmont, first appeared in the 1919 Gourmont issue of *Little Review;* in 1920, it appeared in *Instigations*, as a companion piece to a long essay on Henry James, whom Pound saw as the antithesis of Gourmont, that is, a dispassionate observer in contrast to the active and sensuous Frenchman—though both men represented European "civilization." Years later, in a letter to Katue Kitasono, 29 October 1940, Pound reflected on his own early misappropriation of Gourmont's suggestion: "'Conquérer [sic] l'Amérique n'est pas sans doute votre seul but!' Funny trick of memory. I thought he had written 'civilizer [sic] l'Amérique.' That must have been in my note to him" (L, 347). Pound's memory seems indeed to have failed him, though one does not know what he has forgotten, and there is no recourse to an original intention or document—whether Gourmont's or Pound's. In any case, in 1915, he had quoted Gourmont as saying "Conquérir l'Américain," while the later citation has not only "civiliser" but also "l'Amérique." Pound fails to note the latter misremembrance or revision, which changes Gourmont's goal from a nationalistic aggression to a cultural one and narrows the target from a general "Americanness" to the country somewhat inappropriately called "America," a vague or mythic place, larger than the United States or the continent of North America.

3. SP, 416–17.

4. Untitled review of Richard Aldington's two-volume translation of Gourmont's selected writings, *Dial* 86 (January 1929): 71. Emphasis added.

5. "Prologue" to *Kora in Hell: Improvisations,* in *Imaginations* (New York: New Directions, 1970), p. 26. Hereafter cited in text as Imag.

6. Pound himself drew these connections in his various acknowledgments of Gourmont's influence. Richard Sieburth, in *Instigations,* correctly draws important thematic parallels between Gourmont's writings on Dante and Provençal poetry and specific passages in Pound's early criticism.

7. Kenneth Burke, *Counter-Statement* (New York: Harcourt, Brace, 1931), p. 27.

8. Pound distinguished Gourmont from the more popular Symbolists on the basis of his critical acumen and pedagogical utility: "There are few enough people on this stupid little island who knew anything beyond Verlaine or Beaudelaire [sic]—neither of whom is the least use, pedagogically, I mean. Whereas Gautier and de Gourmont carry forward the art itself, and the only way one can imitate them is by making more profound your knowledge of the very marrow of art" (L, 23). In this letter to Harriet Monroe, Pound fails to cite a specific work or to distinguish between poetry and prose, but he uses Gourmont's critical prose to much greater effect than the poetry. John Espey was the first to make this observation, and his remains a seminal study of Gourmont's impact on Pound's early poetry and poetics. See *Ezra Pound's Mauberley* (Berkeley and Los Angeles: University of California Press, 1955); see especially chapter 5, "Physique de l'amour."

9. At the same time that "New Spain," or America, was undergoing military conquest and colonization at the hands of "Old Spain," a land as much steeped in antique passions for religious crusades as in mercantilist or capitalist lust for gold, that same "Old Spain" was conquered by Renaissance Italian poetry, which provided a new form for epics, while the New World provided heroic themes— heroic only because the Spanish monarchs and a growing audience at home were ready to hear tales of riches and a new frontier in which their cultural inferiority complex—their untimely belatedness—could be forgotten. In *The Conquest of America,* trans. Richard Howard (New York: Harper & Row, 1984), Tzvetan Todorov uncovers Columbus's quixotic venture of launching a crusade to recapture Jerusalem for Christianity with the profits from his voyage to China. He hardly expected to discover America, and his many promises of gold, made both to Isabella and to his dispirited sailors, concealed his real mission of bringing Christianity to China and the Holy Land. About the semiotics of misdirected conquest, Todorov asks: "This discovery seems in truth subject to a goal, which is the narrative of the voyage: one might say that Columbus has undertaken it all in order to be able to tell unheard of stories, like Ulysses; but is not a travel narrative itself the point of departure, and not only the point of arrival, of a new voyage? Did not Columbus himself set sail because he had read Marco Polo's narrative?" (p. 13). Like Ulysses, Columbus returned to a different world, and so did Cortez, who entered upon the scene of Boscan and Navagero. While Todorov does not address the return of Cortez on these terms, he gives an extensive reading of Bernal Diaz del Castillo's history of Cortez and Montezuma. Because this historical period and its various histories captured the imaginations of such modern American poets as Williams, Crane, and Olson, it should receive further consideration. Parallels might be drawn between various trips back and forth between the United States and Europe by such modern Americans as Henry Adams, Eliot, and Pound. And one would want to consider how, after the appearance of English translations of accounts of Columbus's and Cortez's voyages in the 1920s and 1930s, Williams and others took up the Spanish as an imaginative alternative to the Puritan conquest of America—especially in Williams's *In the*

American Grain, which is also an answer to Pound's Whitmanesque and/or Jeffersonian America.

10. William Carlos Williams, *I Wanted to Write a Poem: The Autobiography of the Works of a Poet* (New York: New Directions, 1978), p. 30.

11. The best example of Williams's affirmation of Pound's positive, American influence is "Al Pound Stein," where, yoking the works of the two philosophical and personal antagonists, he says: "The presentation on the New York stage of Miss Stein's *Four Saints in Three Acts,* and the appearance of Pound's Canto XXXVII in *Poetry: A Magazine of Verse,* constitute a dual event of such importance that were our teaching places of any account whatever the nation and the authorities would declare an immediate holiday in their departments of languages until their students had thoroughly familiarized themselves with the works in question." *Selected Essays of William Carlos Williams* (New York: New Directions, 1969), p. 162. Thus, Williams credits Pound with revolutionizing the American language and, along with Stein, making it analytic. He extends Pound's project of reforming the literature departments and curricula at *American* universities.

12. Paul de Man, *Blindness and Insight: Essays on the Rhetoric of Contemporary Criticism* (New York: Oxford University Press, 1971), p. 161.

13. Concurrently, Williams was submitting to the *Dial* a number of "proto-Objectivist" poems, if you will, including "Portrait of a Lady," "Willow Poem," and "Spring Storm." These defiantly modern and American poems vied for space with poems from Pound's *Mauberley,* especially such *French* poems, even "Poems," as "*E.P. Ode pour l'Election de Son Sepulchre.*" More to the point, the "Preface" first appeared in *Little Review* in 1919, one issue after, and apparently postponed for, the Gourmont special issue.

14. *Little Review* V, nos. 10–11 (February/March 1919): 30.

15. *Ibid.,* p. 31. Pound makes a similar, if more conciliatory, declaration in the little poem on Whitman, "A Pact" (*Lustra,* 1916). But even there the United States is a "half-savage country." The background and continuation, if not the genealogy, of Pound's love-hate relationship with the American bard is developed in the Postscript herein.

16. Remy de Gourmont, "Herbert Spencer," *Epilogues: Reflexions sur la vie, 1902–1904,* 3d series, 3 vols. (Paris: Mercure de France, 1905), pp. 244–45.

17. Such figures as the "*bricoleur*" and the "*flâneur,*" the "shuffler" or "amateur" and the "stroller," who wander through the texts and cities collecting ideas, have a long history in French literature and philosophy. Montaigne was a sort of *bricoleur;* the most obvious literary examples are Bouvard and Pecuchet, who also reflect the modern writer's task of disrupting "accepted ideas." More recently, Derrida has associated the figure of the *bricoleur* with Lévi-Strauss, the self-reflexive anthropologist who positioned himself somewhere between the roles of *bricoleur* and engineer and whose reflections on the art of anthropology offer a "myth of mythology," at least by Derrida's account. See *Of Grammatology,* pp. 103–9. Walter Benjamin links the *flâneur,* especially Baudelaire as writer-spectator in the Paris streets, with Poe's "Man of the Crowd" as symbols of the

isolation of the writer in the modern city, but this writer also weaves together the fragments of modern life into poems and stories which "read" modernism. See Benjamin's *Charles Baudelaire: A Lyric Poet in the Age of High Capitalism.* Benjamin, like Nietzsche (a German displaced in Paris) and Derrida, would be given fuller consideration in a study of how "French thought" and "the intellectual" have been (mis-)translated into American letters.

18. Ignored by most critics, these fragments are nevertheless a major part of Pound's efforts to help the *Dial* civilize America. In fact, the *Dial* letterhead of the early 1920s, on which Pound wrote any number of letters, advertised that "during 1920–21 the inedited writings of REMY DE GOURMONT" would appear in that magazine. Gourmont's name was printed in bold capital letters at least ten times larger than the small capitals used for such stellar contributors as Valéry, Croce, Proust, Joyce, and Pound. Moreover, *Dial* had planned to publish much more of "Pound's Gourmont" than ever appeared. When editor Scofield Thayer and Pound had a falling out, the projected translations were abandoned. Still, Gourmont was clearly a drawing card or, by Williams's account, a distraction from American criticism.

19. *Dial* 69 (September 1920): 219. Emphasis added.

20. *Ibid.*, pp. 222–23. Emphasis added.

21. In a forty-one-page preface to *ABC of Reading,* which was part of the never-published *To* Publishers edition (Zukofsky's publishing project obviously challenged the publishing trade by its pre-positional name *To,* which is an address and a challenge), Pound explains the apparent "disappearance" of Gourmont, whose works nevertheless serve as "levers" and "fulcra" to more popular books: "Gourmont's *Le livre des masques* does its work and goes into the archives/ The levers and fulcra used to get certain books thru the press, all this belongs to the records; to be looked up and verified now and then; but has no need to be kept in circulation or considered as READING MATTER for a new generation." "Notes Toward a Preface," Beinecke, p. 30. Pound continued to recommend Gourmont's books to the many poets to whom he sent reading lists.

22. In current usage, "entropy" denotes the amount of randomness or disorder ("random noise") in any system conducting energy or information. In its thermodynamic usage, coined in the nineteenth century by a German physicist, it meant "transformation-content of a system" (*Verwandlunginhalt*), or the fixing of transformational thermal energy into useful work (OED). The definition of entropy became more specifically a wearing down (das Wärmegewicht); to zero energy or stasis. But the more general definition still stands. In sum, the greater the entropy, the less available energy. This is almost literally a winding down, as it was a process generalized to the cooling of the sun and the growing stasis of the solar system. The second law of thermodynamics holds that in any entropic change, the value is either positive entropy or zero—in other words, a loss of energy or, at best, a maintenance. The end of this law or teleological fiction is "universal entropy," the nihilistic vision of nineteenth-century science. Further, as Derrida has suggested, in "White Mythology," and Harold Bloom in *Poetry and Repression,* "en-tropy," the absence of linguistic tropes, became an impor-

tant theme in aesthetics and epistemology at about the same time the thermo-dynamic theories were advanced in geology and physics. Moreover, both the OED and *Webster's Collegiate* cross-reference "entropy," the physical principle, and "trope," or figure of speech.

23. Several critics have noted, yet left unexplored, Gourmont's debt to Nietzsche. Kenneth Burke's cryptic remark is typical: "[Gourmont] recognized that Nietzsche was one of the most important moralists of the time," and further, "there is Nietzsche in every sentence that he wrote," *Counter-Statement*, p. 27. Burke's own method of "perspective by incongruity," which he described as "verbal 'atom cracking,'" owes something to Gourmont's "dissociation."

24. Adams's most striking and ironic treatment of the congruent paths of art, science, and anarchy is "The Grammar of Science," *The Education* (New York: Modern Library, 1931), pp. 449–61, where he notes that such French skeptics as Rabelais and Montaigne had recognized anarchy and multiplicity as universal laws long before modern physics and mathematics (especially the Curies' discovery of radium) had abolished the Darwinian model of unilinear evolution and the hope of biology as a "unified field" or mastering discourse.

25. Remy de Gourmont, "La Morte de Nietzsche," *Epilogues: reflexions sur la vie, 1899–1901*, 2d series, 3 vols. (Paris: Mercure de France, 1904), Vol. II, p. 186.

26. See Chapter One, note 22.

27. "La Morte de Nietzsche," p. 187.

28. "Herbert Spencer," p. 246.

29. *Ibid.*, p. 248.

30. Neither Pound nor Gourmont investigated the long history of science's "poisoning" by literature and writing in general. Nevertheless, Spencer's text teases Gourmont out of scientism into thoughts about writing, thereby restaging the dangerous poisonous/curative effect that Phaedrus's writings about the *pharmakon* performed upon Socrates in Plato's dialogues. I think of a particular passage in Derrida's essay, "Plato's Pharmacy": "Socrates compares the written texts Phaedrus has brought along to a drug (*pharmakon*). This *pharmakon*, this 'medicine,' this philter, which acts as both remedy and poison, already introduces itself into the body of the discourse with all its ambivalence. . . . The *pharmakon* would be a *substance* . . . if we didn't eventually have to come to recognize it as an antisubstance itself: that which resists any philosopheme, indefinitely exceeding its bounds as nonidentity, nonessence, nonsubstance; granting philosophy by the very fact that inexhaustible adversity of what funds it and the indefinite absence of what founds it," *Disseminations*, p. 70.

31. "Herbert Spencer," p. 249. In his own exposé of the "moral origin" of epistemology and all metaphysics, in *Will to Power*, Nietzsche positions himself within a sort of museum, or magic castle, which is his own perspectival "self," equipped with many windows that change the appearance of Truth: "Deeply mistrustful of the dogmas-epistemology, I loved to look now out of this window, now out of that. I guarded against settling down with any of these dogmas, considered them harmful—and, finally: is it likely that a tool is *able* to criticize its

own fitness?" (WP, 221). This passage bears comparison with Heidegger's "Dwelling Building Thinking" and with Gourmont's various houses and museums. We should recall that in *The Case of Wagner,* treated here in Chapter One, Nietzsche parodied Wagner's baroque taste by likening his operas to large, cluttered palaces with huge frescoes comprised of precious bric-a-brac.

32. Writing of *Bouvard and Pecuchet,* Eugenio Donato suggests the havoc museums played with the nineteenth-century dream of adequate taxonomical representation. He makes the crucial distinction between "human knowledge" conceived on the model of the encyclopedic *library* and that of the irreducibly heterogeneous and metonymic *museum:* "The ideology that governs the *Museum* in the nineteenth century and down to the present has often been equated with that of the *Library,* namely, to give by the ordered display of selected artifacts a total representation of human reality and history. Museums are taken to exist only inasmuch as they can erase the heterogeneity of the objects displayed in their cases, and it is only the hypothesis of the possibility of homogenizing the diversity of various artifacts that makes them possible in the first place." "The Museum's Furnace," *Textual Strategies: Perspectives in Post-Structuralist Criticism,* ed. Josue V. Harari (Ithaca, N.Y.: Cornell University Press, 1979), p. 221. If Gourmont and Pound equate museums and libraries, it is to privilege their heterogeneity or failed homogeneity.

33. *Descriptive Sociology, or Groups of Sociological Facts Classified and Arranged by Herbert Spencer,* compiled by David Duncan, Richard Sheppig, and James Collier, 8 vols. (London and Edinburgh: Williams and Norgate, 1873–1881), Vol. I, p. 54. Spencer's criteria for scientific acceptability are the utility of terms and the coherence of detail to theory; he finds these things in his favorite traditional philosophers and scientists and, thus, just catalogues citations.

34. Linking Pound, Nietzsche, and Derrida with a pun on *"seme"* and "semen" and "Dis*seme*ination," Riddel has shown that Pound's "seminal brain" is a trope of trope, suggesting that "turn of mind" that produces "images." See " 'Neo-Nietzschean Clatter': Speculation and the Modernist Poetic Image," cited here in Chapter One, note 10.

35. Rascoe edited the second edition of Pound's translation of *The Natural Philosophy of Love* (New York: Liveright, 1932).

36. Aldington says that "the theory of the transvaluation of values must have been very welcome to a man who delighted to find as many sides of a question as possible." Further, he credits Spencer with providing Gourmont the "intellectual constant," in the form of the quest for a synthetic method, that underwrites his various literary and scientific researches. *Remy de Gourmont,* pp. 8, 15. Such skirting of issues and taking a writer's own words at face value—in this case, Gourmont's idealization of Spencer—are typical of Aldington's assessments of his contemporaries, including Pound, Eliot, and Gourmont.

37. Maurice Blanchot, "The Limits of Experience: Nihilism," *The New Nietzsche,* p. 126.

38. In a fourteen-page unpublished essay, Pound reflects on Gourmont's notion of instinct, linking this with a "capacity to act." This passage is rather ob-

258 Notes, pp. 153–162

scure, but it reflects the conjunction of interests and sources Pound developed in *Guide to Kulchur:* "One's final judgement is 'intuitive'? Or shall I say one's final judgement is made up of a certain number of formulatable reasons and a certain penumbra of imponderabilia. Everything that I write on this subject must be taken with the context of Gourmont's Physique de l'amour and of Fenollosa's essay on the Chinese written character. In Gourmont's exposition instinct is not something opposed to intellect. Intellect is a sort of imperfect forerunner. After the intellect has worked on a thing long enough the knowledge becomes faculty. There is immediate perception or capacity to act instead of a mass of ratiocination." "Further Exposition," unpublished essay, Beinecke, p. 13.

39. For a fascinating account of the overtaking of science by metaphysics, and of both by psychoanalysis and linguistics, see Michel Serres, *Hermes: Literature, Science, Philosophy,* trans. Josue V. Harari and David F. Bell (Baltimore: Johns Hopkins University Press, 1982). Especially suggestive is his analysis of the resistance by mathematics and biology of thermodynamics in chapter 7, "The Origin of Language: Biology, Information Theory and Thermodynamics," pp. 71–83.

40. For a reading of a paradigmatic, self-reflexive, genealogical search for the roots of language in speech, see Jacques Derrida on Lévi-Strauss's *Tristes Tropiques,* in *Of Grammatology,* pp. 101–40. This search for origins, which in the nineteenth-century takes an ironic turn toward linguistics and meta-science, is also Adorno's focus in *Against Epistemology,* trans. Willis Domingo (Cambridge, Mass.: MIT Press, 1983), p. 38, where, not unlike Derrida, he treats the Preface to Hegel's *Phenomenology* as an unintentional undoing of the originary and the new: "With the concept of the first also collapses that of the absolutely new in which phenomenology participated without really coming up with any new themes and so phantasmagorically. The first and the absolutely new are complementary, and dialectical thought had to dispose of (*sich entäussern*) both of them. Whoever refuses obedience to the jurisdiction of philosophy of origins has, since the Preface to Hegel's *Phenomenology,* known the mediacy of the new as well as that of the old."

41. Upon reading such a claim, resting as it does on hyperbolic phallocentrism yet imbedded in an argument that reverses the passive/"feminine" role of reader and that of the active writer, one cannot ignore Pound's confusion. Indeed, even Gaudier-Brzeska's sculpture, "the hieratic head," as Pound termed it, seems here to justify Horace Brodsky and Sophie Brzeska's claims that its obvious phallic shape mocks Pound's sexism and his high opinion of his own powers. For an account of the interpretations of the satiric thrust of Gaudier-Brzeska's statue, see Materer, *Vortex,* pp. 63–70, and see H. S. Ede, *Savage Messiah: Gaudier Brzeska* (New York: Literary Guild, 1931), for a documentary account, from Brzeska's perspective, of the Vortex years.

42. "'Neo-Nietzschean Clatter,'" p. 220.

43. This argument has again gained currency in America through Harold Bloom's revision of the Vichian notion that art and a certain mythopoesis have a premium on Truth, which philosophy, because it rejects the figurative and mystical aspects of language, cannot attain. Bloom inscribes a nearly mystical tropology to the

embarrassment of philosophy and, he hopes, to that of other more rigorous rhetorical (or deconstructive) critics. See especially his reading of Vico addressed to Derrida, *Poetry and Repression* (New Haven: Yale University Press, 1976), pp. 1–21. But such attacks on philosophy can take a far different turn. Derrida, Foucault, and Deleuze are hardly alone in taking different turns back through the history of philosophy. To the list of neo-Nietzchean critiques of the nostalgic search for an adequate, accurate, or even poetic language should be added Adorno's account of the return of repressed figuration in the text of philosophy— especially in *Against Epistemology* and *In Search of Wagner*.

44. Recently, a historian of science, Ben Barker-Benfield, made the point that "the spermatic economy" was a commonplace of the nineteenth-century American eugenics movement. Nearly every textbook for young boys addressed such issues as the relationship between reading and masturbation, the production of sperm and the study of poetry. In contrast to Pound, these manuals treated the spermatic economy as closed and hygienically controllable. Ideally, which is to say moralistically, young boys were to be kept from the very excesses and imbalances Pound encouraged under the name of creative thought. These themes remain to be explored in greater detail in the work of Pound and his contemporaries, who frequently equate male physiology and creativity. Benfield's work is a suggestive beginning, especially "The Spermatic Economy: A Nineteenth Century View of Sexuality," *The American Family in Social-Historical Perspective,* ed. Michael Gordon (New York: St. Martin's 1978), pp. 336–78. In "Border Lines," a note to the translator which runs the length of the essay it presumably explains, "Living On," Derrida at once practices and defines "disseminative translations": "By making manifest the limits of the prevalent concept of translation (I do not say of translatability in general), we touch on multiple problems said to be of 'method,' of reading and teaching. The line that I seek to recognize within translatability, between two translations, one governed by the classical model of transportable univocality or of formulizable polysemia, and the other, which goes over into dissemination—this line also passes between the critical and the deconstructive." "Living On/Border Lines," *Deconstruction and Criticism* (New York: Seabury Press, 1979), p. 93. "Polysemy" and "dissemination" bear an etymological connection to the notion of spermatic economy— and should be explored with regard to Pound's *deconstructive translation*. After making my own discovery of Ben Barker-Benfield, I discovered that my colleague, Joel Porte, had already cited him, to make a connection between economics, reading, and masturbation in textbooks and religious tracts of the nineteenth century. See Joel Porte, *Representative Man: Ralph Waldo Emerson in His Time* (New York: Oxford University Press, 1979), pp. 257–61.

45. *Twilight of the Idols* or *Götzen-Dämmerung* (1889), the last book Nietzsche saw through publication, is a subtle parody of Wagner's apocalyptic vision and his *canonization*—even deification—by the Wagnerites. "*Götzen,*" which means idols or graven images, suggests the direction of his argument, but even more telling is Nietzsche's self-reassessment. For example, he claims that the terms "Dionysian" and "Apollonian" were his first transvaluation and that his own

Dionysianism dates from *The Birth of Tragedy,* a text which he claims was pro-Wagner and anti-Wagner from its inception (p. 110). Along with *Ecce Homo* and *Will to Power,* this text provides Nietzsche's most compelling deconstruction of modernism and "philosophical art." Here he connects God and grammar, which together, as *Logos,* form the philosophical episteme, if not *Darstellung,* that he uncovers in ontology and epistemology alike: "Nothing, in fact, has hitherto had a more direct power of persuasion than the error of being as it was formulated by, for example, the Eleatics: for every word, every sentence we utter speaks in its favour!—Even the opponents of the Eleatics were still subject to the seductive influence of their concept of being: Democritus among others, when he invented his *atom.* . . . [Nietzsche's ellipsis?] 'Reason' in language: oh what a deceitful old woman! I fear we are not getting rid of god because we still believe in grammar" (TI, 38). Frobenius, following Spengler, if not Nietzsche, similarly questioned the role of language in early Greek philosophy and modern science. More of this discussion appears in Chapter Five.

Chapter Five

1. GK, 98. "Gorgias" (which Pound elsewhere misspells "Georgias") makes another appearance in *Guide to Kulchur,* presumably to inaugurate an artistic, if not a rhetorical, revolution against Aristotle's science and logic: "In a sense the philosophic orbit of the occident is already defined, *European thought was to continue in a species of cycle of crisis:* grin and bear it; enjoy life; *variants of Gorgias' dadaism.* Or as they say: Originality of speculative research (guess work) was exhausted with Arry Stotl" (GK, 120; emphasis added). Pound's deliberate attacks are difficult to separate from his simple ignorance of the philosophical tradition. Why does he fail to mention, for example, that Gorgias advanced the arts of rhetoric and poetics against Platonic dialectics? That argument was, to use Duchamp's word, "readymade" in the history of philosophy.

2. Pound, letter to W. H. D. Rouse, May 1937 (L, 295–96). Rouse was founder and general editor of the Loeb Classics about which Pound had a great many questions. While he objected to Rouse's choice of texts for translation and offered suggestions for different interpretations (especially of Homer and Ovid), he approved the basic idea of a bilingual "crib." Moreover, he and Rouse exchanged numerous friendly letters about their parallel attempts to capture the rhythm of "spoken Greek." That they came to different conclusions is more than obvious. The passage I have cited nearly repeats the topics of the Vorticist Manifestoes of 1914, yet Pound proposed that an essay addressing these might serve as introduction or advertisement for the Loeb series.

3. GK, 201. This character can be properly translated as "middle." Pound uses it to suggest neither moderation nor the classical Golden Mean but a transformative conjunction or *translative moment* somewhere between the languages and cultures of China and Western Europe, not a central idea but a series of "self-interfering" cruxes, as Kenner would say.

4. "Further Speculations," Beinecke, p. 1.

5. Even—especially—in his Fascist propaganda, Pound continues his assault upon academic prose and, more radically, upon the English sentence. Against his own earlier and better judgment, he accepts Mussolini's rhetoric precisely because it consists of words divorced from precise meaning and ornaments that defied normal semantics and syntax—"rhetoric," that is, as persuasion. Thus his own sentences are almost necessarily twisted against themselves, not monological and syllogistic, but heterodox and self-effacing. For example: "The DUCE sits in Rome calling five hundred bluffs (or thereabouts) every morning. Some bright lad might present him to our glorious fatherland under the title of MUSSOLINI DEBUNKER. An acute critic tells me I shall never learn to write for the public because I insist on citing other books. How the deuce is one to avoid it? Several ideas occurred to humanity before I bought a portable typewriter" (J/M, 35). Yet Mussolini is more than a repetition of the heroes out of Pound's favorite books; he is master stylist of the sort of mass communication Pound had previously disdained: "Jefferson as a lawyer and as a law scholar used legalities and legal phrases as IMPLEMENTS, Mussolini as an ex-editor uses oratory, and by comparison with Italian habits of speech ('these damned Eyetalyan intellexshuls that think they are still contemporaries of Metastasio'), that oratory is worth study" (J/M, 65).

6. *Ibid.*, p. 2. One does not know whether Pound's figure of half a million is an over- or underestimate. Clearly, the unpublished outweighs the published prose in volume, but the value of these fragments remains open to question. And this openness questions the means by which literary criticism makes its value judgments.

7. At the same time that he refused to tame his own prose, Pound continued to support anthologies that mainstream journals and houses wouldn't publish— or so he claimed, since Eliot did in fact publish several of his recommended authors at Faber. *Active Anthology* (London: Faber, 1933), for instance, included works of Williams, Zukofsky, Marianne Moore, Basil Bunting, Louis Aragon, Eliot, and Pound himself. His *Prefatio,* one should recall, was directed against Eliot's *position* (his place in literary society more than ideas he represented) and against the literary establishment that was already including American and contemporary poetry. He mocks Eliot's conservatism, his inaction, thus: "I am moreover confining my selection to poems Britain has not accepted and in the main that the British literary bureaucracy does NOT want to have printed in England." Then he turns to *correct* Eliot: "If the past 30 years have a meaning, that meaning is not very apparent in Mr. Eliot's condescensions to the demands of British serial publication. If it means anything it means a distinct reduction in the BULK of past literature that the future will carry. I should have no right to attack England's most accurate critic were it not in the hope of something better, if not in England, at least somewhere in space and time" ("Prefatio Aut Cimicium Tumulus," SP, 389, 394).

8. In "Spengler after the Decline," *Prisms,* trans. Samuel and Shierry Weber

(Cambridge, Mass.: MIT Press, 1981), Adorno attributes Spengler's fall from popular hero to ignored cynic to his refusal to see *Kultur* as a reversal of the decline of civilization.

9. Pound approves Levy-Bruhl's—and Leo Frobenius's—anthropological research for precisely those literary and psychological "projections" of the West onto primitive societies which Jacques Derrida has termed "bricolage" and the "myth of mythology." See Chapter Four, note 17.

10. In Renaissance England, *The Nicomachean Ethics,* along with English courtbooks employing its format of rules bolstered by positive and negative examples such as Machiavelli's *The Prince* and Thomas Elyot's *The Book Named the Governor,* was used as Pound proposes to use *The Analects* against Aristotle. In his effort to renew Western philosophy and Confucianism, he returns to old readings, even those contemporaneous with Malatesta.

11. In *Guide to Kulchur* Pound says he will commit himself to as many positions as possible. This jibes with the Fascist panache of *Jefferson and/or Mussolini,* where he claims he'd be a cad not to exercise the freedom of expression that his marginality to the literary establishment afforded him. Williams alludes to Pound's defiance in *Paterson.*

12. Most likely by sheer coincidence, Pound virtually repeats Poe's play upon Aristotle's name and philosophy—Aries Tottle/Total—at the opening of *Eureka,* where Poe irreverently reduces the *total wars* between induction and deduction, idealism and empiricism, theory and eudemonics, into a deflation of serious or philosophical cosmologies. See especially "Eureka," *The Science Fiction of Edgar Allen Poe,* ed. Harold Blower (New York: Penguin, 1976), pp. 213–16.

13. Apparently following Frobenius, Pound equates the earliest African graphics with Dante's use of the vulgate and his own recurrent stress on the verbal or kinetic, as opposed to the nominal and static. He does not so much privilege the originary as claim that as far back as origins can be pushed, the inherent action of language disturbs a simple grammar of voice and things. Even the African tribesmen can be enlisted in the Vorticist movement, however belatedly. See Gaudier-Brzeska's "Vortex," published twenty-five years before the *Guide,* for the interfiling of the Paleolithic and the avant-garde (GB, 20–23).

14. Pound is hardly alone in noting the uses and abuses the category "culture" was undergoing around him. A word that has always seemed foreign to American ears (especially when Whitman felt compelled to answer the growing claims of British culture made by Carlyle), its changed definition and status after Arnold is treated in Raymond Williams, "One Hundred Years of Culture and Anarchy," *Problems in Materialism and Culture* (London: Verso, 1980). And, in a Nietzschean fashion, Adorno, in *Jargon of Authenticity,* trans. Knut Tarnowsky and Fredric Will (Evanston, Ill.: Northwestern University Press, 1973), shows how the privilege afforded German national culture crept into the lexicon of phenomenology.

15. The first American edition of Pound's *Guide* appeared as *Culture* that is, spelled correctly and with a subtitle that identifies it as an autobiography, which could suggest either one of his most scandalous definitions of Kulchur ("What

a man remembers after he has forgotten all he set out to learn") and Coleridge's *Biographia Literaria*.

16. This is the first line of Emerson's *Nature*. The theme of a new literature is often articulated upon old texts, by way of allusion and quotation: not even the call for originality is original or originary. Emerson's insistence on a new, unmediated literature (clearly an oxymoron) was early on reinscribed into Whitman's borrowed revolutionary stance. Another striking instance is Gertrude Stein's opening of *The Making of Americans*, where she seems to cite both Virgil and Marinetti: "Once an angry man dragged his father along the ground through his own orchard. 'Stop!' cried the groaning old man at last, 'Stop! I did not drag my father beyond this tree.'" *The Making of Americans* (West Glover, Vt.: Something Else Press, 1972), p. 3. In this fashion Americans have dragged the New back into the Old World to start over—again—but not like Aeneas or Columbus either.

17. This exhibit, held at the New York Museum of Modern Art in 1934, was an important part of Pound's "paideuma" in the 1930s. He saw such juxtapositioning of the avant-garde and the primitive as proof of a certain structural relationship between things as diverse as "cinematic and modern mechanical" art and the earliest artifacts of human civilization. For Iris Barry's part in the New York connection, see Chapter Three, note 24.

18. *Leo Frobenius 1873–1973: An Anthology*, p. 29. It is difficult to know which Frobenian texts Pound consulted. Presumably, he drew his definition of "paideuma" from *Paideuma: Umrisse Einer Kultur und Seelenlehre* (1921), the short programmatic work which appears nearly in its entirety in this anthology. Hereafter cited in the text as Frob.

19. Pound's notion of the organic is opposed to the organic model Eliot built from Coleridge's ideal of organic art and Arnold's of organic society. The clearest opposition between Pound and Eliot is the opposition of paideuma and palimpsest—or of Dionysian abundance in Canto II—to the metaphor of metaphor of the catalyst in "Tradition and the Individual Talent." Following Gilles Deleuze and Felix Guattari, in the Postscript, I treat Pound's model as "rhizomatic," against the other, "arborescent" organicism.

20. For an exhaustive, German philosophical study of the Platonic dialogues as a microcosm of Plato's ideal of culture, or *paideia*, see Werner Jaeger, *Paideia: The Ideals of Greek Culture*, trans. Gilbert Highet (Oxford: Basel Blackwell, 1944).

21. The pun on "ex-*pli*-cate" (also "com-*pli*-cate" and "im-*pli*-cate") is implied throughout Pound's text, in, for example, Canto IV's "ply over ply." In the *Guide*, he comments on the Greek meaning of the word: "Give the greeks points on explanatory elaborations. The explicitness, that is literally the unfoldedness, may be registered better in the greek syntax" (GK, 279).

22. In a sense, Pound took Frobenius's anthropology to be an archeology in the sense of Foucault's diggings in the various historical epochs for the episteme, the micropolitical and sociological conditions defining systems of power and control. If Pound hardly approached the rigor of Frobenius, who hoped to read the

causes of tribal behavior in reactions to geographical and cultural requirements, he nevertheless did hope to discover motives in layers of dead facts. Curiously enough, Foucault also uses the figure of a moving "palimpsest" for the "archeological" layers of written history. See Foucault, "The Prose of the World," *The Order of Things,* pp. 17–42. For a literary critical treatment of palimpsest, see Gérard Genette, "Proust Palimpseste," *Figure I* (Paris: Editions du Seuil, 1966), p. 67, where Baudelaire's *Les Paradis l'artificiels,* a text woven into *The Cantos,* is cited as a palimpsest which reflects on palimpsests. In "The Double Session," *Disseminations,* trans. Barbara Johnson (Chicago: University of Chicago Press, 1981), Derrida addresses palimpsest and Genette's taming of this exemplary figure that literature has used both to bind and to free metaphoricity and intertextuality. See especially pp. 248–52.

23. The German notion of *Kultur* had undergone at least two nationalistic or *reichsdeutsch* transformations since Nietzsche's time, that is, during the time of the two World Wars. Because Frobenius, Burckhardt, and Spengler at once criticize and subscribe to the fundamental superiority of the German culture and "race," it becomes difficult to place them in what I have schematized in Chapter One as the Nietzschean or Wagnerian interpretations of modernism and the tradition. Pound rejected the word *Kultur* because of its associations with National Socialism as well as its relation to the sort of linguistic and literary research he everywhere criticizes as "abstract" and "compartmentalized."

24. Jacob Burckhardt, *Force and Freedom: Reflections on History,* ed. James Hastings Nichols (New York: Pantheon Books, 1943), p. 93. Much earlier Pound had captured some of this sense of Burckhardt in his use of the electrical metaphor (for a force at once untamed and determining) and his selection of such luminous details/electric facts from Burckhardt's *Renaissance,* "I Gather the Limbs of Osiris," SP, 22.

25. *Ibid.,* p. 94.

26. In a rather incoherent and perhaps incomplete letter to Wyndham Lewis, 29 October 1936, to which he attached his "Manifesto against TREASON of the Clerks," Pound complains that Spengler's undeserved popularity is due to the fact that Frobenius, much more the "active thinker" or "factive personality," was untranslated. This and other letters, which suggest a closer relationship between the late careers of Lewis and Pound than scholars have noted, are held at the Olin Library, Cornell University, Ithaca, New York.

27. Oswald Spengler, *The Decline of the West,* 2 vols. in 1, trans. Charles Francis Atkinson (London: George Allen & Unwin, 1932), p. 18. Hereafter cited in text as Decline.

28. Adorno contends that, in such late works as "The Physiognomy of the Modern Metropolis," Spengler predicted Goebbels's media manipulation and Hitler's Caesarism. Spengler was decidedly no friend of the masses, but his utter refusal to disguise *Realpolitik* in the redemptive power of *Volk* or *Kultur* put him out of favor with the National Socialists. See Adorno, "Spengler after the Decline," *Prisms,* pp. 53–72. There might be a slight parallel with Pound's own

refusal to endorse the nationalistic thrust of Anglo-American culture between the wars. As Fredric Jameson suggests, in his book on Wyndham Lewis, the proto-Fascism of more than one modern writer arose from disgust at nationalism, Western European economic and other declines.

29. When he edited *Guide to Kulchur,* Eliot deleted fifteen libelous pages. I know of only six unexpurgated copies of the book, one of which is held, as part of the remnants of Pound's Rapallo collection, at Humanities Research Center, University of Texas, Austin. These pages of "topical" references and scatalogical epithets add little to Pound's serious criticism, but one should recognize them as part of his "paideuma," not to say one of his strategies against Eliot's more decorous writings.

30. In 1934 Pound added a Postscript to *Gaudier-Brzeska* in which he recalls the devastating effects of World War I on European culture. He took the death of the young sculptor Henri Gaudier-Brzeska as a symbol of the dissolution of modernism into a series of new nationalistic arts. This Postscript collapses the metaphors of vortex and paideuma, assigning both the function of producing "whirling" or "revolutionary" resistance to what Spengler had called the decline of Western culture as a whole.

31. This hangover from Rousseau as well as German Romanticism, by which primitive culture was thought figurative, abundant, poetic, and active—as against mechanistic—persists in Heidegger (not to mention several contemporary Americans, including Bloom) and is revised by Derrida and Adorno, as cited above.

32. This concern reenters "The Pisan Cantos," where Pound mentions Rouse's discovery that the pronunciation of Odysseus and Elias was so nearly homonymic that the legends of epic hero and biblical prophet were confused. To this confusion, Pound adds the names OY TIE, which means no man and is the name by which Odysseus identified himself to Cyclops; and Ouan Jin, a character in one of Frobenius's African tales, whose mouth had to be stopped because his words created too many things. Out of this palimpsest of puns and self-reflections, and in characteristic fashion, Pound begins *The Cantos* over again at LXXIV: 425–27.

33. As I have suggested, Pound's readings in the Chinese language are interested and creative. He seems to apply a notion of "grinding" or analysis to individual characters, very few of which lend themselves to such picture reading. This character does not contain an element which means "mortar," and, frankly, it is difficult to imagine one here. Instead, as Professor Shuhsi Kao has informed me, the bottom mark can mean "son"—though no reader of Chinese would recognize this as a separate part of the character unless he were playing or punning across several languages and cultures.

34. Pound is wrong in claiming to be the first to so honor Heraclitus. Nietzsche also "had the crust" to praise the presumed author of "Everything flows" in a similar, if more rigorous fashion. References to Heraclitus dot has later work, especially *Twilight of the Idols,* where he writes: "I set apart with high reverence the name of *Heraclitus.* When the rest of the philosopher crowd rejected the

evidence of the senses because they showed plurality and change, he rejected their evidence because they showed things as if they possessed duration and unity" (TI, 36).

35. As early as *Spirit of Romance,* Pound noted the intersections of heterodoxy and orthodoxy, polytheism with Christian monotheism, and the diminished imaginative force of the latter. "Axiomata" (1909), reminiscent of Blake's "Proverbs of Hell" and Nietzsche's *Zarathustra,* provides the clearest instance of his preference for gods over God: "The greatest tyrannies have arisen from the dogma that the *theos* is one, or that there is a unity above the various strata of theos which imposes its will upon the sub-strata, and thence upon human individuals" (SP, 51).

36. Again, Pound's preference for Heraclitus over Aristotle—he might have said Socrates—affiliates him with Nietzsche, who cites Socratic *ressentiment* as proof of the decadence of Greek philosophy: "It is not only the admitted dissoluteness and anarchy of his instincts which indicates *decadence* in Socrates: the superfaction of the logical and that *barbed* malice which distinguishes him also points in that direction" (TI, 30). Pound begins to suggest such a critique when he complains that Aristotle spoiled the "guess work" and other enjoyments of philosophy.

37. "Incipit B," Beinecke prose. This rather garbled fragment contains notes toward definitions of several key terms.

38. Michel Foucault, *The Order of Things,* p. xvii. Foucault's analyses, in this book and elsewhere, of the nineteenth-century episteme bear consideration with Pound's involvement in the new sciences. Especially relevant are his efforts, in *Archaeology of Knowledge,* to place aberrant texts like those of Nietzsche and Jules Michelet in a "corpus," then a "system," then the "episteme"—efforts which have been parodied by Derrida. Foucault approaches his task of reading through textual strata with a mixture of scholarly rigor and Nietzschean irony, but his irony fails him when he must somehow organize the fragmentary observations Michelet called history. Pound had a rather different notion of Michelet's contribution: "The modern historian lives after Michelet, and Michelet was right in at least one thing, that is his perception of different strata in modern Europe and with different predispositions and habits differentiated per strata, but functioning in accord with specific desires." "On Historians," Beinecke, p. 1. While Pound's notion of "desire" might better be plotted as "Le Voyage Gastronomique" than, say, an "archeology of nineteenth-century discourses," his writings might be thought to add an interesting turn to such readings. I mean only to sketch the outlines that further study of Pound's criticism of history and historians would have to take.

39. "Notes Toward a Preface," Beinecke prose, p. 4.

40. See "Hors Livre: Outwork," Jacques Derrida's Preface to *Disseminations.*

41. Paul de Man, *Allegories,* p. 115.

42. Invoking both Pound's name and his notion of the "prospective," in an essay entitled "Projective Verse," Charles Olson makes the following distinction between the projective and the English (or Eliotic) verse traditions: "(projectile

(percussive (prospective *vs* the NON-Projective (or what a French critic calls 'closed' verse, that verse which print bred and which is pretty much what we have had in English & American, and have still got despite the work of Pound and Williams." *Selected Writings,* ed. Robert Creeley (New York: New Directions, 1966), p. 15. Consideration of this countertradition, based largely on those "opinions" which Pound hoped would evade an academic, if not a *poetic,* interpretation, goes beyond the scope of this study. Still, as my Postscript suggests, Williams, Olson, and Whitman would appear prominently in a comprehensive study of Pound's Kulchur. In fact, they appreciated the com-*pli*-cations of Pound's text with a subtlety matching that of current rhetorical (and mostly French) critics.

43. In the context of Canto LXXIV, "hilaritas" refers to classroom playfulness over the scanning of Sappho's poetry. But the word also refers to Christ's perfect beatitude. In the *Guide,* Pound comments on an Italian textbook's use of "the hilarity of thy face" to describe Christ: "These seem to me to belong rather to the universal religion of all men, than to any sect or fad of religion" (GK, 141–42). Thus, what is apparently Pound's ignorance of a Christian attribute becomes, through his familiar tropology, a Classical reference, a statement about a certain lighthearted familiarity with Homer, and a joke, told perhaps by Homer's laughing gods at the expense of Pound.

Postscript

1. Pound's habit of grounding his poetics, of erecting his own poems, on difficult, overused, or abused metaphors is properly termed catachresis, which for Aristotle and Cicero meant far-fetched metaphor as well as a coinage for something that already had a proper name. Another name, and thus a catachresis, for this "troping on trope" is *metalepsis,* the mastering trope which Harold Bloom says is "the only trope reversing trope . . . produces the illusion of having fathered one's own fathers," Harold Bloom, *Poetry and Repression,* p. 20. Despite Pound's insistent concerns about canonization and revision, Bloom does not admit this prodigal son of Emerson and Whitman into his new canon of "The American Sublime," perhaps because Pound's own tamperings with the American tradition forbid the isolation of stable tropes and clear Oedipal crossings, or perhaps it is because Pound uncovers a certain reactionary poetics ingrained in America's poetic revisionism. See Chapter One, note 15, for Derrida's opposing treatment of catachresis.

2. In "La Loi du genre/The Law of Genre, *Glyph 7* (Baltimore: Johns Hopkins University Press, 1980), pp. 176–232, Jacques Derrida takes advantage of the puns in the French word *genre* (gender, literary genre, law, generation, and genealogy), which are not automatically available to speakers of English or American. Yet the manner in which sexuality, the *conception* of ideas, literary lines, and aesthetic laws converge in the production and reproduction of texts is also an American theme (in, for example, the "Calamus," "Children of Adam," and "Sea Drift" sections of *Leaves of Grass;* Henry Adams's figures of the Virgin and the Dynamo; and Pound's fascination with Provençal mysteries and with

Remy de Gourmont's *Physique de l'amour*). This is not to equate philosophical rigor—be it Foucault's or Derrida's—with a certain poetic reflexivity.

3. "Teufelsdrock," a chapter of *The Education*, recounts the founding of this contradictory party in 1901 by Adams and several of his fellow idlers. In a hilarious string of parodic allusions, Adams plays fast and loose with such writers as Schopenhauer, Hegel, Kropotkin, and Carlyle, in order to "satisfy man's need to destroy by reaching for the largest synthesis in its [philosophy's] ultimate contradiction." Henry Adams, *The Education of Henry Adams*, p. 407. Subsequent references to this book are in text, abbreviated as E with appropriate page numbers. Likewise, I will refer to *Mont-Saint-Michel and Chartres* (Boston: Houghton Mifflin, 1905); abbreviated in text as *MSMC*.

4. As important a reader as Harold Bloom, the reigning exemplar of American revisionism, virtually repeats F. O. Matthiessen's canonization of Emerson and Whitman in *The American Renaissance* (1941). Matthiessen notes the Emersonian imperative to begin culture over again, and it should be remembered that he also comments on Whitman's interest in a uniquely American rhetoric and pedagogy.

5. This and other Whitman citations are from *Complete Poetry and Selected Prose*, ed. James E. Miller, Jr. (Boston: Houghton Mifflin, Riverside Editions, 1959). Hereinafter cited in text as LG (*Leaves of Grass*) and DV (*Democratic Vistas*), with page numbers of this edition.

6. Marcel Schwob was one of the many names Pound picked up in his readings of Remy de Gourmont. Pound never explains his praise of the marginal novelist-historian who is remembered for his fragmentary and largely fictionalized accounts of the Children's Crusades, which were organized by Eleanor of Aquitaine, one of Pound's heroines. Since Pound's many allusions to the Crusades fail to mention Schwob, the fact that Schwob is a model historian for the genealogically complex, "nomadic," and "rhizomatic" character of his *La crusade des enfants* (1896) might be little more than a coincidence. Yet such coincidences or accidents of textual affiliation are precisely crucial questions for modern and contemporary writing—in poetry as well as science. Regarding Schwob's self-reflexive history, see Gilles Deleuze and Felix Guattari, "Rhizome," *On the Line*, trans. John Johnston (New York: *Semiotext(e)* "Foreign Agent Series," 1983), p. 55. Hereinafter cited in text as OL.

7. "Rhizome," an excellent initiation into Deleuze and Guattari's strategies, appears as the introduction to their second volume of writings on the micropolitics of "schizo-culture," *Mille plateaux: capitalisme et schizophrénie* (Paris: Minuit, 1980). It also appeared separately as *Rhizome: Introduction* (Paris: Minuit, 1976), from which Johnston made the translation I have used.

8. This letter, quoted by Williams in "Prologue to *Kora in Hell*," shows that, notwithstanding his own efforts to bring language "closer to the thing," Williams was shameless about quotations, even those about quotation. Pound indicts this questionable American habit in "Patria Mia" (1911): "Nine out of every ten Americans have sold their souls for a quotation. They have wrapped themselves about a formula of words instead of about their own centres" (SP, 102). One

can perhaps detect a faint echo of Emerson's metaphor of the circle with its center and circumference. For a thorough study of Emerson's and Whitman's similarly contradictory positions on quotation and originality, see Joseph Kronick, *American Poetics of History, From Emerson to the Moderns* (Baton Rouge: Louisiana State University Press, 1984).

9. Deleuze and Guattari do not explain the American undoing of the "book." Here one might cite Derrida's examination of the undoing of philosophy's foundations by literary waywardness, which rests upon the opposition of texts to books or writing (*écriture*), to speech. As is well known, Whitman often meditates on the structure of *Leaves of Grass*, and, of course, he gives no final reading of the figures of "the book [of Nature?]" or of his "leaves" or leavings or, to use the Poundian phrase out of context, "loose-leaf method." Perhaps most relevant here are such provisional statements as the line from "Shut not your doors to me proud libraries . . . " *Inscriptions:* "The words of my book nothing/the drift of it everything" (p. 13); and, as adumbration of Pound's semantic and syntactic disruptions in the name of the "ideogram," see ll.1–9, "Poem #6," *Song of Myself:* "A child said *What is the grass?* . . . Or I guess it is a uniform hieroglyphic" (p. 28).

10. *The Journals and Miscellaneous Notebooks of Ralph Waldo Emerson*, ed. Merton M. Fealts, Jr. (Cambridge, Mass.: Harvard University Press, 1973), vol. 10 (1847–1848) p. 79. In a book obsessed by genealogy, Melville uses the same figure of wild nature for Pierre's misreading of his own apparently stable American lineage: "Still are there things in the visible world, over which shifting Nature hath not so unbounded a sway. The grass is annually changed; but the limbs of the oak, for a long term of years, defy that annual decree. And if in America the vast mass of families are as the blades of grass, yet some few there are that stand as the oak; which, instead of decaying, annually puts forth new branches; whereby Time, instead of subtracting, is made to capitulate into a multiple virtue," *Pierre, or the Ambiguities,* eds. Harrison Hayford, Hershel Parker, and G. Thomas Tansell (Evanston, Ill.: Northwestern University Press, 1971), p. 9. In Melville's late collection of poems, *Weeds and Wildings*, this tangle becomes titular and textual—as it is for many American writers.

11. Critics generally skip over The Jefferson/Adams Cantos, which are little more than random and undigested misquotations from the presidential letters. Such doubled sins of omission are well documented by Peter Shaw, "Ezra Pound on American History," *Partisan Review* 44, no. 1. (1977): 112–24. Shaw fails to see that Pound's abuse of America's "democratic history," and its de facto forgiveness by his apologists, is as damaging to Pound's Fascist "order" as it seems motivated by it. For a more developed treatment of the connections between modern writing and antidemocratic thought, see Fredric Jameson, *Fables of Aggression: Wyndham Lewis, The Modernist as Fascist.* (Berkeley: University of California Press, 1979).

12. These "inrooted ideas" are elsewhere figured as "ply over ply" of allusions (IV: 15, *passim*), which is another indirect and overdetermined quotation—that is, a catachresis. Critics have found Pound's source in "pli selon pli," from Mal-

larmé's "autre Eventail," where it refers to the opening and closing of a fan; from Browning's *Sordello,* where it refers to an architectural ornament; it also refers to a Chinese poem by Li Po. The critics agree, however, that it is a (dis-)organizing metaphor for *The Cantos.* Such characteristic textual metaphors do locate Pound's poem in a tradition—a marginal tradition of self-reflexive or textually obsessive writings that he traced far back into antiquity.

13. The connection between Brooks Adams and Pound is detailed by Earle Davis in *Vision Fugitive: Ezra Pound and Economics* (Lawrence, Ks.: University of Kansas Press, 1968). For a fuller, critical reading of Brooks Adams's theories and metaphorics of entropy and economics, see Kronick, as cited in note 10 above, especially pp. 167–69. Andrew Parker provides an informed, important, and anything but merely celebratory account of Pound's economics in "Ezra Pound and the 'Economy' of Anti-Semitism" *Boundary 2,* nos. 1, 2 (Fall–Winter 1982–83): 103–28.

14. Pound's references to Henry Adams are generally short condemnations of the historian as the entropic "wearing down" of the Adams family line. Though he seems to have overlooked the relevance of Adams's work to his own, Pound's very dismissals are couched in *The Education*'s figure of history as a "dynamo." From what he felt was Adams's overaestheticized nostalgia, Pound wants to save "the possibilities of a revival, starting perhaps with a valorization of our cultural heritage, not merely as something lost in dim retrospect, a tombstone, tastily carved, whereon to shed *dry tears or upon which to lay a few withered violets in the manner of . . . Henry Adams. . . .* [but] *'As monument'* or I should prefer to say a still workable dynamo left us from the real period nothing surpasses the Jefferson correspondence" (SP, 147; emphasis added). Building thus on metaphors, on an admixture of monumental history and living things, Pound carries on the work of the American genealogists, of whom Adams is but one of the more self-conscious.

15. Perhaps working with Pound makes one more accutely aware of the dangers of making the simple choice between fragmentation and order. Pound's choices were not as simple as they might appear, certainly not so simple as to excuse his dismissal on the grounds that he does not bear careful study within the American canon, and in the context of those modernisms that did not end so simply in Fascism and/or madness. The words of Erich Auerbach, written during his exile from Nazi Germany, assume a new poignancy at the present time when quite different simplifications threaten, many offered in the name of social responsibility and, to use Adorno's phrase, "the jargon of authenticity." I quote Auerbach, who seems for a moment to echo Adams, if not the other American modernists I have considered: "So the complicated process of dissolution which led to fragmentation of the exterior action, to reflection of consciousness, and to stratification of time seems to be tending toward a very simple solution. Perhaps it will be too simple to please those who, despite all its dangers and catastrophes, admire and love our epoch for the sake of its abundance of life and the incomparable historical vantage point which it affords." *Mimesis,* trans. Willard R. Trask (Princeton, N.J.: Princeton University Press, 1953), p. 552–53.

INDEX

Adams, Brooks, 149, 225, 270n
Adams, Henry, 139, 212, 221, 226–31, 253n, 270n; *The Education of Henry Adams,* 212, 215, 230, 268n; "The Grammar of Science," in *Education,* 256n; *Mont-Saint-Michel and Chartres,* 212, 227–31
Adams, John, 216, 218, 224
Adorno, Theodor, 26, 265n, 270n; *Against Epistemology,* 258n, 258–59n; *In Search of Wagner,* 240n, 259n; *The Jargon of Authenticity,* 262n; "Spengler after the Decline," 261–62n, 264n
aesthetics as system, 4–5, 41, 238n; Hegel's, 165; Heidegger on Nietzsche's, 15, 68; Nietzsche's attack on, 17, 20, 26, 28. *See also* System
Agassiz, Louis, 61, 75; "Essay on Classification," 62; *Recherches sur les poissons fossiles,* 62
Aldington, Richard, 151, 252n, 257n; *Remy de Gourmont,* 237n
American renaissance. *See* American risorgimento
American risorgimento, 51, 80, 211, 214
"Anaximander's Fragment," 73
Anschauung, 108
anthropomorphism, Gourmont's use of, 155–56, 162
aporia, 39, 62
Aquinas, St. Thomas, 32, 42, 106
Aristotle, 32, 39–47, 63, 70, 171–73, 190–92, 196–200, 244n, 262n, 266n; on rhetoric, 41–42; Pound's attacks on, 197; *The Art of Rhetoric,* 41–44; *The Eudaemonic Ethics,* 173; *Logic,* 199; *The Nicomachean Ethics,* 172–73, 196, 200, 262n; *Poetics,* 41–43, 46, 244n
Arnold, Matthew, 13, 55, 103, 111, 217, 263n
Auerbach, Erich, 270n
Aufhebung, 17. *See also* Hegel
Augustine, St., *Confessions,* 245n

author/authority, 109
author/authorship, 220
autotelic, 37, 90, 227

Babbitt, Irving, 84
Barker-Benfield, Benjamin, 259n
baroque gallery, 144, 226
Barry, Iris, 178, 250n, 263n
Baudelaire, Charles, 12, 140, 155, 264n
Bayreuth, performances of Wagner's operas, 17, 21–22, 30
Beinecke Library, Yale University, Ezra Pound Collection, 16, 240n, 242n, 245n, 247n, 249–51nn, 255n, 258n, 261n, 266n
Being, 15; Heidegger on, 7; question of, 8; vs. Will to Power, 68
Benda, Julien, *La Trahison des clercs,* 129
Benjamin, Walter, 39, 243n, 249n, 254n; *Charles Baudelaire: A Lyric Poet in the Age of High Capitalism,* 250n, 255n; "The Task of the Translator," 245n
Bernstein, Michael, *Tale of the Tribe: Ezra Pound and the Modern Verse Epic,* 82
Blanchot, Maurice, 151, 257n
Blast, 93
Bloom, Harold, 238n, 248n, 255n, 258n, 268n; *Poetry and Repression,* 259n, 267n
book, as opposed to text, 16
Borges, Jorge Luis, 200–201
Born, Bertran de, 229; *Dompna Soiseubuda,* 50
Boscan, and Navagero, 130–31
Bouvard and Pecuchet, 79, 127, 254n, 257n
bracketing, 7, 9, 235n. *See also* Heidegger; Husserl
bricolage, 262n
bricoleur, 136, 185, 254n
British Museum, as metaphor, 71, 146